I0003155

Sometimes...

Punch Press Optimization, the Travelling Salesman Problem, Isohodes,
and the GREAT QUESTION

CASE STUDIES IN COMPUTER SCIENCE
VOLUME 3

Λεγει αυτω Ιησους: εγω ειμι 'η 'οδος.
Dicit ei Iesus: Ego sum via.
Jesus said to him: I am the Way.

– John 14:6

"Somebody has to do the hard jobs."
— Mark Weaver in *The Wreck of the* Phosploion

In this case, the problem isn't the *problem*. The problem is the *solution*.
— James M. Waclawik

I revert to the doctrinal methods of the thirteenth century,
inspired by the general hope of getting something done.
— G. K. Chesterton
Heretics CW1:46

"I often stare at windows."
— G. K. Chesterton
"The Crime of Gabriel Gale"
in *The Poet and the Lunatics*

* * *

There are two ways of getting home; and one of them is to stay there.
The other is to walk round the whole world till we come back to the
same place...
— G. K. Chesterton
The Everlasting Man CW2:143

* * *

For more information, visit
http://DeBellisStellarum/CSCS/cscs.htm

* * *

Case Studies in Computer Science

Volume	Title
0	The Problem with "Problem-Solving Skills"
planned	(on the theory of strings and molecular biology)
planned	(on local ad insertion for cable TV)
3	Sometimes...

Sometimes...

Punch Press Optimization,
the Travelling Salesman Problem,
Isohodes,
and the GREAT QUESTION

Peter J. Floriani, Ph.D.

CASE STUDIES IN COMPUTER SCIENCE
VOLUME 3

The Habitation of Chimham Publishing Company
Titusville, Florida

Dedication

Ad Majorem Dei Gloriam
Ἐτοιμασατε την ʽοδον κυριου, ευθειας ποιετε τας τριβους αυτου.
Parate viam Domini, rectas facite semitas eius.
Prepare the way of the Lord, make straight His paths.

Matthew 3:3 citing Isaiah 40:3

With special thanks to
Jim Waclawik, M.S.
Wayne Stahre
Leo Schwartz
Nancy Carpentier Brown
Rita M. Floriani

In Memoriam
my parents
my teachers
my co-workers
Samuel L. Gulden, M.S.
Samuel R. Frankel, M.S., P.E.
Charles H. Blouch

Text and images
Copyright ©2016 by Peter J. Floriani, Ph.D.
All Rights Reserved

Cover art-road PSDGraphics
ID:24 105691585461
74100455257170070054698

ISBN-13: 978-0-9899696-3-5
Library of Congress Control Number: 2015953166
Library of Congress Catalog-in-Publication-Data

CONTENTS

SERIES FOREWORD   IV

FOREWORD FOR THIS VOLUME   V

CHAPTER 1: THE PUNCH PRESS AND OPTIMIZATION   1

CHAPTER 2: THE TRAVELLING SALESMAN PROBLEM   10

CHAPTER 4: ABOUT THE GREAT QUESTION   51

CHAPTER 5: THE END OF OUR PRESENT PATH   88

APPENDIX 1: ABOUT ISOHODE BOUNDARIES   89

APPENDIX 2: THE SHEARING PROBLEM   97

APPENDIX 3: PSEUDO-CODE   98

APPENDIX 4: STIRLING NUMBERS OF THE FIRST KIND   107

BIBLIOGRAPHY   108

This series of monographs is my attempt to enrich your own personal collection of previously solved problems – which is, in the end, the only "problem-solving skill" worthy of the name.

The "monograph" format is not the one usually expected in the modern academic world. People expect journal articles, or perhaps FACEBOOK postings. However, the monograph is a traditional approach to exotic topics for many disciplines: even Sherlock Holmes said (in *The Sign of the Four*) he was "guilty of several monographs." Besides, it is gratifying to explore such fascinating topics by this means, thereby aiding in the advance of Science writ large, and in the pursuit of Wisdom.

> I can imagine Sherlock Holmes remarking, in a light allusive fashion, that he himself had written a little monograph on the subject of cows' tails; with diagrams and tables solving the great traditional problem of how many cows' tails would reach the moon; a subject of extraordinary interest to moonlighters. And I can still more easily imagine him saying afterwards, having resumed the pipe and dressing-gown of Baker Street, "A remarkable little problem, Watson. In some of its features it was perhaps more singular than any you have been good enough to report. I do not think that even the Tooting Trouser-Stretching Mystery, or the singular little affair of the Radium Toothpick, offered more strange and sensational developments."
>
> GKC, *Irish Impressions* CW20

* * *

At the Ambrosian, we are constantly striving to synthesize – that is to unite – the various disparate topics and subjects of knowledge – indeed, we strive to see, and therefore to carry out all that is implicit in seeing, that most scientific phrase in the Creed: *Per quem omnia facta sunt* = "through Him all things were made."

We desire to be Christians first: followers of Jesus Christ, and therefore will arm ourselves with every possible weapon in the war we must wage until our deaths. That means we call upon science and literature, upon mathematics and philosophy, upon history, language, engineering... it is said in many ways, and shall be said in many ways, but for us, there is no such thing as a different subject, simply because we wish to say with St. Paul, we have resolved to know nothing but Jesus Christ and Him crucified. (1 Cor 2:2)

However, that does not mean we are locked into some sort of a morbid perpetual Good Friday asceticism. What it means is that we try to always be conscious of what it is we are doing. We have our Final End in mind, whether we are studying or reading or lecturing or working in a laboratory or playing a game or an instrument, or even walking across the campus. That is why our bells ring every hour: to remind us how little time is left to us.

– from the introduction to the Course Catalog of the Ambrosian University

FOREWORD FOR THIS VOLUME

I was not going to do this volume now. I was going to get two others done first, on the wild monoid and on papally assisted file transport, hoping that I would run out of time before having to risk everything on such a bizarre presentation as you will find in this book. But when not one, not two, but *three* of my friends urged me to get my work down on paper, just in case... well. Besides, they said, *sometimes* somebody has to ask even *stupid* questions, because that's how real science may advance.

The scary thing is that not that I asked a stupid question.

The scary thing is that I found a completely unexpected answer.[1]

But then the question has been brewing for a long time. Indeed, it is one of the longest-standing questions in my own experience. (I will not comment here on how long it has puzzled others.) As you will hear shortly, it was about 35 years ago, in 1978, soon after the new HP3000 time-sharing minicomputer was installed at Frankel Engineering Labs, that I spent some time on the very interesting question of optimizing the path for a numerically controlled (NC) punch press. At that point in time, I was a young hotshot programmer, and I had never had any classes on complexity or algorithms, and so I coded as I still often do, "by the seat of my pants." That's foolhardy if not dangerous, but it can get things done, and *sometimes* one has very little choice in the techniques at one's disposal.[2]

Of course the situation at FEL hardly permitted the academic approach of quiet contemplation and careful research, long hours in the queues and stacks of the library, driving (with whip and scourge) a gang of tired grad students to distill countless irrelevant journal articles, perhaps the convocation of seminars or colloquiums of eminent scholars to debate the matter, or inquisitive letters to authorities in the field (that was long before e-mail). That company didn't have a large library of reference works, and though the technical collection at the local public library was *far* better back then than it is now, it did not have any texts relevant to such an esoteric issue.

This is not an *apologia* for my choices, or even for the solution you are about to learn, though you will get to hear about that in this text, and some more about the underlying problem. *Sometimes*, we ought to revisit even classic problems just for the sake of the exploration – that is, for the sheer adventure[3] of the thing. We *may* learn something new: it's the way in which science has always worked.

Peter J. Floriani, Ph.D.

---

1   Or so it appeared to me. But then I've read Hayes's *Project: Genius* which tells how a young man set out to show that the Earth was round, only to discover it was flat. Then again, it may be an instance of what my friend Wayne Stahre calls the "Fosbury Flop."

2   Of course this recalls GKC's famous line: "I revert to the doctrinal methods of the thirteenth century, inspired by the general hope of getting something done." (*Heretics*, CW1:46). GKC is hinting at theoretical foundations such as *purpose* (not a particular form of worship) and that is critical, as you will hear.

3   Adventure: "A hazardous and striking enterprise, a bold undertaking in which hazards are to be met and issue hangs upon unforeseen events." *Black's Law Dictionary*, 72

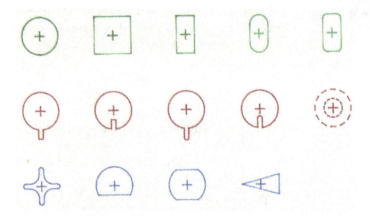

Plot-program forms of punch press tools
(as drawn by Frankel Engineering software)

First row: round, square, rectangle, obround, rectangle with rounded corners.
Second row: four different keyways, knock-out.
Third row: radius-corner, dee, double dee, wedge.

One of the more useful things which occurred during my employment at Frankel Engineering Labs in Reading, Pennsylvania from 1977 to 1983 was my introduction to INDUSTRY: its methods, its tools, its issues and problems and special cases, and its people.

This introduction began on my very first day, with my immersion into the mysteries of Numerical Control programming. You can read about my experience on that marvellous day, September 19, 1977, in the first volume of this series, so I need not repeat the story here. But NC programming, the arranging of instructions for machine tools, is not the same sort of thing as regular computer programming. To keep them distinct we refer to the machine tool kind as "part-programming," and we often use the term "software development" for the computer kind. Part-programming is quite different from software development, simpler in some ways and far harder. It is that specialized skill needed in order to take a blueprint which describes the geometric and metallurgical character of a desired item, and arrange the "instructions" which will control a particular machine tool to produce that item as specified.

Now, there are various kinds of these machine tools. There are lathes which spin cylinders of metal and cut them into shafts. There are milling machines of various kinds from simple two-axis types to very complex multiple axis types used in production of aeronautical parts. There are drills which may be considered just a limited form of a milling machine. Lathes, mills, and drills take advantage of rotation: either the cutting tool or the part itself spins as the metal is cut. There are flame cutters and other such tools which use various advances in physics to cut the metal; some of these are quite amazing and can accomplish tricks which are very hard to attain with a rotating tool.

There are also punch presses, which work on sheet metal.

A punch press uses a tool called a *punch-and-die* which is a kind of heavy-duty cookie cutter together with a kind of socket which matches its outline. Pairs of these are arranged in a large rotating wheel called the *turret*, so that any particular tool (punch-and-die) can be positioned under the working head of the machine tool. There, the power of hydraulics is applied, so that when a sheet of steel is placed between the punch-and-die at a desired coordinate, the head slams together the punch and the die, chopping out a little steel cookie. That sheet is held in clamps and shifted back and forth, and the turret revolves as necessary to bring the proper tool under the head, and so eventually that sheet of steel is nibbled and gnawed and pounded so that it attains the desired arrangement of holes as specified on the blueprint.

All sorts of fascinating issues in mechanics, drafting, metallurgy, and the detailed craft of metal-working and machining arise in such a task. For example, there is the very trivial issue of the setting the position of the clamps which hold the piece of sheet metal. These *workholders* must be placed so that no punches are made in the safety zone near them. (Think of the old tree-trimming rule, "Don't cut off the branch you're sitting on.") Of course if the blueprint requires holes along both top and bottom edges, arranging those workholders will involve some trickery.

But the chief of these issues is the "shape" of the punch-and-die, that is, the geometric figure cut into the metal by each particular tool. There are four major shapes in frequent use: rectangles, squares, rounds, and "obrounds" which are a rectangle to which semicircles have been attached on each end. There are also keyways, dees, double-dees, knock-outs, radius-corners, and more specialized shapes used in certain applications.

All these tools come in a variety of sizes – the rounds, for example, ranged from as small as 1/16 inch (0.0625") in diameter to 4 inches. Certain tools such as rectangles and obrounds could be set either tangentially or radially relative to the turret:

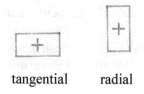

tangential   radial

That is, they may be mounted so that their long dimension is parallel to either the X axis or the Y axis of motion, in order to nibble out slots longer than the tool. There are also adapters to enable the punch-and-die to be set at an arbitrary angle.

Aspects such as these are important, since part of the engineering work we part-programmers had to accomplish was the checking of our work using a plot program. This piece of software read in the finished work in the form of a text file, together with a list of the tools (the punch-and-die shapes) installed in the turret and other data such as the blank size, and generated a drawing which would either be displayed on the screen or drawn on a plotter. This diagram was drawn to scale and showed all of the punches which would be made, thus permitting the checking of the result against the blueprint.

Several technically interesting challenges also arise. For example, when the sheet of metal is to be bent after being punched, the dimensions on the blueprint are typically specified relative to the final product, not the flat sheet. This requires the computation of a "bend allowance" to handle the fact that there is a certain "stretching" of the metal; all dimensions relative to the bends must be adjusted by a small amount.

Another issue occurs when the sheet is larger than the working area of the machine tool: that is, when the blueprint requires holes at dimensions larger than the punch press can move. This requires a special trick called "repositioning," in which the machine grows an extra pair of hands to hold the sheet of metal while the workholders shift backward, thus permitting the punching of holes in a new area of the sheet.[4]

These issues were, of course, quite common in actual part-programming, and were not particularly difficult to handle. Plotting was very easy. We augmented the part-program with "pseudo-code" instructions, written as "comments" in the punch-press language, or otherwise arranged for.[5] These instructions specified the few critical matters which were not part of the part-program:

1. The size of the blank (unpunched sheet).
2. The location of the workholders.
3. The current tools in the various stations of the turret.

We even had an extensible module for the plot program which enabled the user to add a "tool" of a completely unique shape as necessary.

The purely technical matters required attention, but were rarely a serious challenge. Bend allowances can give some headaches when the part has lots of bends, and one must take every detail into account: Is the bend inwards or outwards? Is it at 90 degrees? Are the dimensions interior or exterior? And *what* is the thickness of the metal being used? – but careful attention to such details is just the usual thing demanded in any kind of programming. Repositioning was another kind of headache, mainly because every machine tool did it differently, and it is quite easy to do it incorrectly – it's rather like flipping the sign of a variable in algebra. But even the idea of doing something like cutting a very large circular hole by chipping away with a big two-inch round, first in a spiral and then nibbling along the circumference... that's just a nice little practical application of trigonometry, and may be left for homework.[6]

---

4   This is one of the techniques used in order to handle the issue brought up earlier, where the blueprint requires holes along all four edges of the sheet. The part is repositioned, not because it is too long to fit under the machine tool, but in order to dodge the safety zones of the work-holders. (Yes, such things occurred in parts we actually made part-programs for.)

5   For example, there was a separate program called PUNCH which produced the paper tape version of the part-program, since paper tape was the form of input then used by most machine tools. PUNCH could be configured so that it would not punch the "comments"; thus the text file (the part-program as stored in the computer) could contain the data required for making a plot, and also be used to produce the paper tape for the machine tool.

6   If you feel like trying it yourself, here is the assignment. Given the radius of a hole to be nibbled, the coordinate of its center, the size of the round to be used, and the maximum "tooth" to be left, compute the angular increment necessary to produce such a hole. There are two variants: (1) the "slug" version, in which a ring is nibbled, and the remaining central disk falls down (the "slug"), or (2) the "complete" version, in which the central disk is nibbled away completely as efficiently as possible, then the outer ring is finished as required. (Yes, some CNC forms of punch press had commands to punch such holes automatically, but not all machine shops had such presses, so we had to make arrangements to perform such chores in our software.)

There is something else, however, which arises in the work of part-programming for a punch press, and when it arose for us at FEL, I had no idea what I was about to tackle. Still less did I envision that some 35 years later I would be working on the problem again, or writing a book about it.

Optimization

We called that problem *optimization*,[7] and though I came to think of that operation in a different way when I learned about compilers, it was the first meaning I learned for this term as a computational task. To put it simply, the problem was about the *order* in which the work was to be done, the order in which the holes were to be punched. But I would rather tell you the little story about how I came to learn about the matter.

In my first volume of Case Studies monographs, I mentioned one of our customers, Reading Sheet Metal, located just a few blocks away from FEL in Reading, Pennsylvania. They had a Wiedemann punch press, and were one of the users of our punch-press packages. They had even bought a plotter so they could do their own checking of their part-programming, and they were very happy with it. Some of their engineers came to our offices one day, sometime in 1978, to discuss making things better, since they knew we were getting a new computer, and they wanted to know how their software might be improved.

The biggest issue, of course, besides making sure that our time-share system stayed up a lot more than the old ones (which liked to crash, sometimes multiple times in a day!) was the question of efficiency. Not so much of the software, though they were very attentive to the cost of using our system – but the effective performance of the part-program which resulted from our software. They expected that the part-program would be correct: that is, it would cut the metal as required by the blueprint. But there also were choices to be made about the *order* of the punching, and this order would dictate the total time which would be consumed in the making of the part.

Our customer waxed eloquent about this. It sounded a little sharp, but the man was trying to be fair to all parties concerned in the matter; to my knowledge he was a good manager, and not a brutal task-master. But the point he raised was significant, and we will have to consider it in justice in order to deal with... with what we must eventually deal with.

He said: "Look. My press operator grabs a blank (the unpunched sheet of metal). He throws it on the bed of the press, gets it into the clamps and aligned. Then he presses the START button and he sits back, *doing nothing and getting paid*, while the press runs the part-program. And when the green light goes off and the clamps open, he grabs the finished part off the press, throws it on the stack of finished parts, and starts all over again. So all the time the press is running that part-program I don't get any work out of him!

---

7   From the Latin adjective *optimus* = best, the superlative of *bonus* = good.

4

So I think you can understand why it's a big concern to me that we get *the very fastest possible performance* out of these part-programs... even if we have to pay a little more computer time."

My friend and co-worker Charlie Blouch and I looked at each other and shrugged. Charlie was a superb engineer and taught me NC, part-programming, and related engineering matters. We turned to our boss, Samuel R. Frankel, a kind and generous man, and a professional engineer who had been applying computers to engineering projects for decades. (In 1960 he had bought a CDC 160, one of the first computers in the Reading area, in order to help a local steel company with thermodynamics problems for their furnaces.) Sam smiled and told our customer, "We'll look into this and get back to you with a solution."

Samuel R. Frankel   Charles H. Blouch

Now, you (and most people who are reading this) probably know all about finding minimum-length Hamiltonian paths, which are the shortest[8] way of visiting a collection of points on the Euclidean 2-plane: that very famous problem called the Travelling Salesman Problem – and that brings up the topic of complexity and the even more famous GREAT QUESTION of computer science, about P versus NP... and you will be wondering why I didn't speak up. But at that early day I had not heard of any of those things. I was simply a young hotshot daredevil programmer, who hadn't discovered there were hard jobs lurking out there, ones which very serious and experienced scholars had banged their heads against without getting anywhere. At least with the new computer I felt I had a shot at getting some sort of result for our customer, since I knew the interpretive BASIC on our old systems was ridiculously slow and severely limited in memory, and our glorious new HP3000 minicomputer, installed in August of 1978, would help me solve their puzzle. (Even better, its FORTRAN and even its BASIC was recursive!)

Now, I won't try to reconstruct the solution I devised; I don't have the code, and I don't recall the details of what I did, though it was probably

---

8   The *shortest* way is not always the *fastest*: consider the brachystochrone curve for a famous counter-example, and a good lesson for us computer scientists.

a straightforward "brute-force" attempt at finding the optimum solution.[9] However, it will be good for us to note one or two interesting aspects of the matter, just because these aspects are an essential part of the problem, and may not be excluded from consideration just because of a desire for theoretical purity or something. Moreover, it is better to consider this as the "punch press optimization problem" and not something abstract in graph theory (like the Travelling Salesman Problem) for the same reason: that is, we were trying to satisfy a particular customer, not establish some sort of theoretical result. This is again an important point, and we shall comment on it shortly. (Recall I said earlier that *Purpose* matters.)

A Statement of the Punch Press Optimization Problem

I will not do this as rigorously as typical in computing; it is not quite as important just now. (Besides, if you are a theory hog you will have your fill later, yes, theorems *and* proofs.)

We are given a part-program for a punch press,[10] which is simply a list of X and Y coordinates together with a given "tool" or turret number which is an index to one of the punch-and-die pairs presently installed in the turret, indicating which of these is to be applied at that given coordinate. For the time being, we ignore the matter of repositioning.

We apply the standard Euclidean distance equation, but we must also be given the velocity for the punch press being used. We must also know the effective time required to index the turret, and the total number of turret stations, in order to handle the change from one tool (punch-and-die pair) to another. Sometimes it will be better to change turret stations, make a hit, and change back, rather than to make a long motion without changing the turret; such actions must be decided as part of the algorithm.

Hence we do *not* have the classical Travelling Salesman Problem here, where the edge weights are strictly Euclidean distances; we have an extended one, with edge weights determined by other factors as well as distance. As we said, these extensions are essential to the customer's need, and the question of finding the optimal order or path to visit all the given coordinates with the appropriate punch-and-die pair in the turret must be solved with all of them taken into account.

We made various attempts at solving this problem. I recall one in par-

9   Every programmer knows it is quite easy to write a program to solve the travelling salesman problem. (Pseudo-code for a "simple" version is provided in Appendix 3.) That's not the hard part. The hard part is getting it to run fast enough on real problems (say for $n>12$) to satisfy the user.

10   In general we consider the NC form, rather than the CNC form, which provided complex instructions to produce regular patterns such as grids, or nibble slots and holes of various shapes. We designed ways of treating such things and while the matter is of some academic interest, it is a digression from the main topic, so we avoid it here. For homework, the reader may examine the theory of the question, then design a method of treating such things. They simply alter the concept of a vertex (which has the same to-coordinate as the from-coordinate) to an entity for which these two are not identical.

ticular which I left running over a weekend. When I came in on Monday they asked me how much longer it would be running.

"It has 495 coordinates, I don't know... maybe another day or two..." I said in my abject ignorance, and youthful hope.

How funny it is to think of this today, more than 37 years later. If that machine still existed and was functional, and nothing else had interfered, that program would still be running, in fact would have barely begun its work. Hard to believe...

But that is not how the story ends. We knew that this attempt at finding the best solution was not suitable, so we tinkered around with ways to get "good" solutions. Eventually, after banging our heads against it for a while, we thought we had something that might be acceptable, and we asked the engineers to come to our offices for a demonstration.

We showed them the program, and the manager shook his head. "Look. You shouldn't doing that big hole first, it makes the sheet wobbly, and then it might get jammed when you go to do that far corner. And then this... these here... we wouldn't do those holes first, we'd catch them on the way up that other column... No, this isn't an optimal solution."

And so, rather to my surprise, I learned one of the most important lessons of my life. All too often the customer does *not* want the "best" solution in the mathematical sense. He wants the one which looks like what *he* would do in solving the problem. That one is the BEST.

In other words, we needed to give the user control over the arrangement of the work, and not expect the software to work out that "best" solution. It was not any sort of theoretical matter, but a user-oriented one.

As a result, we invented the idea of "priorities": a scheme that lets the user group the coordinates for selected holes into several classes, each represented by a small integer. These classes were performed in strict ascending order, regardless of any other "optimality." Then another routine would find a suitably efficient order to handle all of the coordinates in that class, starting with the "nearest" to the previously handled coordinate, though without regard to what would follow. This turned out to be a very good approach, since it also enabled us to handle repositioning in a very straightforward manner: that special action was performed only at a boundary between two priority classes, thereby segregating the coordinates into pre-reposition and post-reposition groups.

What's the Point If You Cheat?

So it sounds like we cheated. Sure, we cheated! We weren't finding the "optimal" solution according to the theoretical definition.[11] But then that

---

11 Don't worry; you will *not* be cheated in what is coming, though it may be not what you expect.

was not what was wanted of us. We weren't asked to solve *that* problem, a theoretical problem, as elegant and interesting as it is! We were asked to solve a different problem, one which had a very specific sort of application, the results of which had to be judged for "effective efficiency" by stern judges of what made sense to them as engineers, as people running a production house, who wanted to get results as effectively and efficiently as possible.

Even if we had been able to supply them with the *guaranteed* "best" solution from the mathematical viewpoint, if that "best" way did not make sense to them, it didn't count. It had to be "best" as they decided.

If it's a cheat, it's because (in retrospect) one *thinks* this is just a variant of the Travelling Salesman Problem, even if it has awkward things like turret-changing added in, and so one expects to find some sort of wacky but cool insight into that most difficult and yet most exciting riddle. So one feels cheated since the punch press optimization problem turned out to be something where the user has to guide the machine to do things in an appropriate order. It's a completely different sort of game, and it feels like the rules were changed.

The rules weren't changed. I *said* that our problem was to make the part-program the way our customer wanted... It's our own fault in not understanding him; he can't be expected to give us an elegant and formal statement of the problem, as if he was a computer scientist! He's running a sheet metal fabrication plant, after all; he just wants to keep his employees busy so their pay is just, and get the products to *his* customers in a timely manner and at a fair cost.

Kind of Like Science, Not Engineering

In that sense, then, one must fall back on science. One cannot simply "engineer up" the solution to something which has not been completely analyzed, and may not be completely understood. What really was the idea of this "optimal" path, anyway? The wise computer scientist (like that certain young man at FEL) has to sit and listen to the customer, because what he (the software developer) thinks is the question being asked might only be his own concept of the riddle, and not what the customer has in mind.[12]

This is why my good friend and fellow computer scientist Jim Waclawik has said "Computing is an experimental science." Often we must propose solutions for a given problem – but we must be ready to discover they are quite unsuitable for use by the customer, or for the actual real-world problems awaiting us.

At the same time, if we *computer* scientists know our *specific* field well, we ought to have at least an awareness and an interest in the *general* field of Science, which means we ought to know something of both its dead-

12 Humility is hardly a word one expects to find in such contexts, but that virtue is one of the hallmarks of a scientist, if he is to have any hope of discovering truths of the Real World. I examine this very difficult matter in another context, that of molecular biology, in the second volume of my Case Studies monographs.

ends and its advances. Hence, speaking as a *scientist* it is well for us to retain some awareness of these attempts, as pathetically useless as they appear at first, since sometimes the truth is hidden in the waste. Consider the work of Marie and Pierre Curie in their pursuit of radium. These audacious and unbelieveably persistent scientists spent from 1898 to 1902 processing a ton (2000 pounds, or about a thousand kilograms) of the residues of pitchblende ores – that is, the waste from a uranium mine. They repeated the same lab chore of fractional crystallization hundreds of times, and in the end had about a tenth of a gram of radium chloride.[13]

Who knows what sort of things might yet lurk in other waste heaps... but it takes wisdom and persistence to find out. But those, along with humility, are the virtues of a scientist, and if we want to have further advances in our pursuit of Truth, we may need to do something quite boring, maybe with what others deem useless waste. We might not find something as splendid as radium, the tiny tubes of which the Curies loved to see glowing in their darkened lab – a difficult and primitive workplace, nothing more than an old shed, unheated and leaky when it rained – and yet we may find something to shed light on real-world problems.[14]

---

13   Glasstone, *Sourcebook on Atomic Energy*, §§5.6-5.9.
14   That is a concentration of ten million (0.1 g Ra obtained from ca. 1,000,000 g of waste). Granted they were not aware of its terribly dangerous radioactivity: Marie Curie died from cancer resulting from her exposure to such highly radioactive substances. That is another sort of cautionary tale about our own work; fortunately very few naturally occurring algorithms are radioactive.

We are going to investigate the famous Travelling Salesman Problem, (often shortened to TSP) which we shall now define.

Our Given Data

We are given a set G of n nodes which are distinct points[15] in the Euclidean 2-plane:

$$G = \{v_0, v_1, \ldots v_{n-1}\}$$

where $v_i = (x_i, y_i)$.

This object G is a *graph* in the mathematical sense: a set of n vertices (also called nodes) and $e = n(n-1)/2$ edges which join every distinct pair of nodes. However, every edge of this graph also has a *weight* which is the Euclidean (Pythagorean) distance between its vertices. The edge between nodes p and q has the weight $dist(p,q)$ defined to be:

$$dist(p,q) = \sqrt{(x_p - x_q)^2 + (y_p - y_q)^2}$$

A Statement of the Problem

Given a graph G as defined above, we wish to find a tour t, defined to be a sequence of integers between 0 and $n-1$,

$$t = <t_0, t_1, \ldots t_{n-1}>$$

where $0 \le t_j < n$, and where $t_i \ne t_j$ for $0 \le i, j < n$. That is, the t_j are indices of the nodes of G, thereby specifying a *permutation* of the points of G.

This tour t indicates a closed path touching every node once; the path returns to its start from the last node.

Moreover, this tour must be such that the sum of distances included on that path is *minimal*. That is, we wish to find the permutation t which minimizes:

$$total(t) = \sum_{i=0}^{n-2} dist(t_i, t_{i+1}) + dist(t_{n-1}, t_0)$$

This problem has been studied at great length for most of the existence of computer science. Clearly it appears to require the consideration of all $n!$ possible permutations or tours, thereby placing this Travelling Salesman Problem in the class of those "hard" problems known as NP, the shorthand for those problems solvable in polynomial time by a non-deterministic machine. Indeed it has been shown to be NP-hard; all this can be found in the

---

15 For ease in handling the theory, we restrict the nodes such that no three are co-linear. If necessary this may readily be arranged in practice by a slight perturbing of the coordinates.

vast literature on complexity and is omitted here. This study approaches the matter from another direction which does not seem to be explored in any of the existing studies, except for an all-too-brief hint in the massive study of the topic.[16]

It is all too easy to feel daunted, even dismayed, at the rapid growth of that misleadingly cheerful factorial, $n!$, which exceeds the range of 32-bit integers (about 4.3e9) between 12 and 13, and exceeds the range of 64-bit integers (about 1.8e19) between 20 and 21. At 24! (about 6.2e23) we are very close to the famous Avogadro's Number of items per mole, from which it's not far to 52! (about 8.1e67), the number of possible shuffles of a standard deck of playing cards. Very dismayed. These are *huge* numbers, and what if one has 495 nodes to handle, which means checking 494!/2 tours (about 4e1117)? Is there *any hope at all* for such cases?

At the same time, it should be noted, one must guard against the over-complexification of any given problem. It is true that the even more common problem known as *sorting* also requires the selection of a suitable sequence from among all possible permutations of the indices into the array of values to be sorted. However, sorting is emphatically *not* a complex algorithm of $O(n!)$ run time! One can easily sort in $O(n^2)$ time, and with just slightly more effort in $O(n \cdot \log n)$ time, and in some situations even faster than that. Such techniques are learned at a very early level in the study of algorithms, and should come as no surprise.

The question which has beset serious scholars for at least half a century is whether there exists any algorithm which might (as in sorting) find the suitable permutation to solve the TSP without having to consider all $(n-1)!/2$ possible tours.

If this was a detective novel I might here drag a red herring across the path, but it is intended to be a technical monograph, so I will spill some of the beans now. But even such a silly word as *beans* may open a profound study of a topic as serious, as interesting, and as debatable as *miracles*.[17]

I am rarely serious, and only pretend to be a scholar on rare occasions. However, I have spent some time in my lab playing with graphs,[18] and I have encountered something very strange, interesting, and debatable: something worth telling you about.

As stunning as it will be for you to read, there appears to be a way. *Sometimes.*

---

16 "...if all points lie on the border of the convex hull then the border itself is an optimal tour." Applegate *et al.*, *The Travelling Salesman Problem*, 33.

17 "I casually summed up the distinction between the supernatural and the unreasonable by the phrase that one might believe that a Beanstalk grew up to the sky without having any doubts about how many beans make five." GKC, ILN May 21, 1910 CW28:531

18 I heard this in grad school from a fellow student: "There's this feeling you get when you play around with graphs: there are a lot of graphs."

Some Basic Details from Graph Theory

Edge Count

Our given graph G of n nodes $n_i = (x_i, y_i)$, points in the Euclidean 2-plane, are numbered beginning with zero. The graph G is *complete*, that is, there are edges between every pair of nodes, and they are assigned a weight given by the above distance equation. Since by combinatorics there are $n \cdot (n-1)/2$ distinct pairs of n nodes, the number of edges e is given by:

$$e = \text{EdgeCount}(G) = \frac{n(n-1)}{2}$$

See the "Convenient Table" below for the values for some small values of n and some other useful related values.

Edge Index and Name

For simplicity in discussion (when our graphs are sufficiently small in size) we identify each of the edges according to a simple arrangement. For computational use we define the "edge index" by a simple relation:

$$\text{EdgeIndex}(i,j) = (1 + i + (j-1)(j)/2)$$

The following table (i is the row, j is the column) shows Edge Index:

from/to	1	2	3	4	5	6
0	1	2	4	7	11	16
1		3	5	8	12	17
2			6	9	13	18
3				10	14	19
4					15	20
5						21

By simply converting these indices into letters, we have edge "names" for study of the results. (For a lower-case representation in ASCII, add 96 to the edge index.)

from/to	1	2	3	4	5	6
0	a	b	d	g	k	p
1		c	e	h	l	q
2			f	i	m	r
3				j	n	s
4					o	t
5						u

Thus, the edge from 2 to 4 is given by $w=(1+2+(3)(4)/2) = 9$, so its name is "i".

12

For example, consider the following graph:

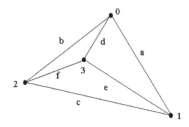

There are four nodes: 0, 1, 2, and 3. There are six edges:
    a joining 0 and 1
    b joining 0 and 2
    c joining 1 and 2
    d joining 0 and 3
    e joining 1 and 3
    f joining 2 and 3.

Tours, Retrograde Tours

A *tour* of a graph G is a possible permutation $t = <t_0, t_1, ... t_{n-1}>$ of the nodes of G which may be proposed as a solution for the TSP of that graph. It contains all n nodes without repetition, and thus it also contains n edges, as it must not be overlooked that there is *one additional edge* linking from the last node of the tour back to its starting node. (That edge returns the travelling salesman to his starting point.)

Also, given a sequence of items
$$z = <z_0, z_1, ... z_k>$$
we define its *retrograde* as that sequence in reverse order
$$z^R = <z_k, z_{k-1}, ... z_1, z_0>$$

Hence, for any tour t we also have its retrograde t^R, which visits the same nodes but exactly backwards.

However, we do not consider two tours distinct when one is simply the reverse of the other, or when they use exactly the same edges and only have different starting nodes. Hence, in the above example, there are only three possible distinct tours:
    1. from node 0 to 1 to 2 to 3 and back, which uses edges a, c, f, d.
    2. from node 0 to 1 to 3 to 2 and back, which uses edges a, e, f, b.
    3. from node 0 to 2 to 1 to 3 and back, which uses edges b, c, e, d.

We note that the nodes in any given tour may be visited either forwards or backwards; that is, we may traverse the edges joining them either forwards or backwards, and any node may serve as a starting point, since we will arrive at it eventually after completing the tour. None of these choices alters the tour, which is the same regardless of the direction we choose, or the node we start at.

Since a tour touches each node exactly twice, we can represent a tour either by its permutation of the nodes themselves, or by the permutation of the indices of the nodes; either form shows the order of their visitation. We can also represent that tour by a list of the edges it traverses, since we can readily derive the order of visiting the nodes from that list. This is sometimes more convenient, and also permits an elegant simplification.

Canonical Tour (Edge form)

We are able to represent a tour by its list of edges in ascending (lexicographic) order. This is a very convenient representation, which we call the *canonical tour*. Since the collection of edges in a tour completely determines the order of visiting the nodes, the order of listing the edges in a tour is therefore irrelevant. The three tours from the above graph therefore have the following canonical representations:

    1. acdf (from node 0 to 1, to 2, to 3, and back to 0)
    2. abef (from node 0 to 1, to 3, to 2, and back to 0)
    3. bcde (from node 0 to 2, to 1, to 3, and back to 0)

Canonical Tour (Node form)

Given a graph G of n nodes (x_i, y_i) numbered from 0 to $n-1$, we define a *canonical* tour (in node form) of G to be a permutation of the n integers from 0 to $n-1$, representing indices of the nodes of G, such that:

    (1) the first index is *always* 0.
    (2) the second index is always *less than* the final index in the series.

Consider a tour t and its retrograde t^R, where $t_0 = 0$. Only one of these is canonical, the one for which $t_1 < t_{n-1}$, that is, the second index is smaller than the last index.[19]

Again considering the above example, we have the following cases:

    1. from 0 to 1 to 2 to 3 and back, or 0, 3, 2, 1
        – but its canonical form is 0123.
    2. from 0 to 1 to 3 to 2 and back, or 0, 2, 3, 1
        – but its canonical form is 0132.
    3. from 0 to 2 to 1 to 3 and back, or 0, 3, 1, 2
        – but its canonical form is 0213.

To conclude this discussion of canonical forms, *any* tour of G has a corresponding canonical tour which visits the nodes by the same edges. That tour may easily be found by cyclically permuting the given tour until the first

19 Note that this technique readily gives the formula for the TourCount for a given graph, which we define next, that is, the number of (canonical) tours, since there are $n-1$ nodes to permute, and only one of the two possible orderings (either t or its retrograde t^R) is canonical.

node of the tour is node 0, then checking to see if the index of the second node of the tour is smaller than the final one. If not, the retrograde of the tour (from its second to its final entry) is performed. For example, given the tour 3,1,0,2, we cycle it until the first node is zero:

3,1,0,2
1,0,2,3
0,2,3,1

But now the last node is *less* than the second one, so we must retrograde it (that is, we walk it backwards, keeping the zero-node first):

0,1,3,2

which is the canonical form of the given tour.

We note that each canonical tour represents a total of $2n$ sequences of nodes; furthermore, all of these have the same canonical edge tour.

For example, the canonical tour 0132 also represents the following seven tours:

1,3,2,0 and 1,0,2,3
3,2,0,1 and 3,1,0,2
2,0,1,3 and 2,3,1,0

and 0,2,3,1.
All of which, however, are not in canonical form.

The Tour Count
The number of possible distinct (and hence canonical) tours for a graph G of n nodes is given by:

$$\text{TourCount(G)} = \frac{(n-1)!}{2}$$

Proof. A tour is simply a permutation of the n nodes of the graph. Without loss of generality, we always start every permutation with the first node, $t_0 = (x_0, y_0)$, thus we only need to consider the permutations of the remaining $(n-1)$ nodes, of which there are $(n-1)!$ possibilities. Furthermore, we note that the direction by which we traverse a tour (forwards or backwards) makes no difference, as they have equal total distances; thus we must divide that number of permutations by two.

The Number of Covering Paths
Sometimes we will need to know how many paths there are through a given set of n nodes, starting and ending at arbitrary nodes and visiting all nodes. This number is given by:

$$\text{CoveringPathCount}(n) = n!$$

Proof. We may choose any of the n nodes as the first point, any of $(n-1)$ nodes as the second point, and so on, which is nothing more than the permutations of n items; by combinatorics this is $n!$. Also in this case, we must consider both going from start to end by a given path, and its exact retrograde path, so we do *not* divide by two.

The Number of Edge Pairs

In a graph G with n nodes and $e = n \cdot (n-1)/2$ edges, the number of possible *pairs* of edges is:

$$\text{EdgePairCount}(G) = e(e-1)/2$$

Proof. We may choose any of the e edges as the first edge, and any of the remaining $(e-1)$ edges as the second, and divide by two since we do not need to distinguish the first from the second. (This is simply the combination of e edges taken two at a time.)

We may also express this in terms of n:

$$\text{EdgePairCount}(G) = (n+1)(n)(n-1)(n-2)/8 = (n^4-n^3-n^2+2n)/8$$

The Number of Tours Containing a Given Edge

In a graph G of n nodes, the number of tours containing any given edge is:

$$\text{ToursContainingEdge}(G) = (n-2)!$$

n	tours	tours with edge
3	1	1
4	3	2
5	12	6
6	60	24
7	360	120
8	2520	720
9	20160	5040

Proof. Since every one of the n nodes occurs on every tour, we may assume that the start point of every tour is the from-node of the given edge. Also, since we may proceed in either direction, we assume we always traverse the given edge *from* that start (its from-node) *to* its end (its to-node). From there, we must arrange a (non-circular) path leading from the to-node of the given edge back to its from-node: this path must include all of the $(n-2)$ remaining nodes. There are $(n-2)!$ permutations of those $(n-2)$ nodes. Hence, there are $(n-2)!$ such paths, all distinct, and therefore there are that many

16

tours containing the given edge. (Note that they are distinct, irrespective of a permutation and its retrograde, since in the first case the tour will include the edge from the given edge's to-node to the *first* node of that particular permutation, but in the retrograde case, the tour includes the edge from the given edge's to-node to the *last* node of that permutation, which is *not* the same edge.)

Tours Which Contain a Pair of Edges

In a graph G of n nodes, the number of tours containing any given pair of edges, ToursContainingEdgePair(G,**a**,**b**), is given by

$(n-3)!$ if the two edges *share* an endpoint
$2(n-3)!$ if the two edges *do not share* an endpoint

n	tours	sharing	nonsharing
4	3	1	2
5	12	2	4
6	60	6	12
7	360	24	48
8	2520	120	240
9	20160	720	1440

Proof. We are given edges **a** and **b**. Without loss of generality, we assume we always start at the from-node of **a**, proceed to the to-node of **a**. After this there are two cases to consider:

Case (a): the second edge shares an endpoint with the first edge. The argument proceeds as when finding the number of tours containing one edge, but here there is one less node to be considered. Hence the number of tours is $(n-3)!$

Case (b): the second edge does not share an endpoint with the first edge. Thus, there is at least one node intervening between the two given edges on every tour, and all such intervening nodes are not endpoints of either of the given edges. Since there are $(n-4)$ such nodes, we have the following two series of possibilities:

First we check
     a-from, **a**-to, **b**-from, **b**-to, (the other $n-4$ nodes)
     a-from, **a**-to, (1 other node) **b**-from, **b**-to, (the other $n-5$ nodes)
     a-from, **a**-to, (2 other nodes) **b**-from, **b**-to, (the other $n-6$ nodes)
     a-from, **a**-to, (3 other nodes) **b**-from, **b**-to, (the other $n-7$ nodes)

17

and so on up to
     a-from, **a**-to, (n–6 other nodes) **b**-from, **b**-to, (the other 2 nodes)
     a-from, **a**-to, (n–5 other nodes) **b**-from, **b**-to, (the other 1 node)
     a-from, **a**-to, (n–4 other nodes) **b**-from, **b**-to
for a total of n–3 cases.

We also check
     a-from, **a**-to, **b**-to, **b**-from, (the other n–4 nodes)
     a-from, **a**-to, (1 other node) **b**-to, **b**-from, (the other n–5 nodes)
     a-from, **a**-to, (2 other nodes) **b**-to, **b**-from, (the other n–6 nodes)
     a-from, **a**-to, (3 other nodes) **b**-to, **b**-from, (the other n–7 nodes)
and so on up to
     a-from, **a**-to, (n–6 other nodes) **b**-to, **b**-from, (the other 2 nodes)
     a-from, **a**-to, (n–5 other nodes) **b**-to, **b**-from, (the other 1 node)
     a-from, **a**-to, (n–4 other nodes) **b**-to, **b**-from
which again is a total of n–3 cases.

Since there are $(n–4)!$ possible permutations of $(n–4)$ nodes, which arrangement we may divide by any of the $2(n–3)$ instances listed above, there are therefore
     $2(n–3)(n–4)!$
possible tours, which is equal to
     $2(n–3)!$
as was stated above.

* * *

All this might have been elementary or even boring to some readers, but it helps (as well as being quite Euclidean) to set forth a little about our notation and ways of talking about these things. Also it's good mental exercise.

n	e	Tours	TwE	Eps	Epn	rC	rCCH
3	3	1				0	0
4	6	3	2	1	2	1	2
5	10	12	6	2	4	5	5
6	15	60	24	6	12	15	9
7	21	360	120	24	48	35	14
8	28	2520	720	120	240	70	20
9	36	20160	5040	720	1440	126	27
10	45	181440	40320	5040	10080	210	35
11	55	1814400	f	f	f	330	44
12	66	19958400	f	f	f	495	54
13	78	239500800	f	f	f	715	65
14	91	3113510400	f	f	f	?	f
15	105	43589145600	f	f	f	?	f

Column heads:

n: the number of nodes (vertices) in the given graph.

e: the number of (undirected) edges. Since every graph is complete,
$e = (n)(n-1)/2$

Tours: number of possible (canonical) tours:
formal value is $(n-1)!/2$

TwE: tours containing an edge: formal value is $(n-2)!$

Eps: tours containing a pair of edges which *share* an endpoint:
formal value is $(n-3)!$

Epn: tours containing a pair of edges which *do not* share an endpoint:
formal value is $2(n-3)!$

rC: observed number of edge-crossings in a regular polygon of n nodes. It appears to follow the formula $n(n-1)(n-2)(n-3)/24$. For n even, the nodes are slightly perturbed so that no three edges meet in a single point. Hence the regular hexagon has 15 crossings, not 13.

rCCH: in a regular polygon, maximum number of chords of convex hull: formal value is $((n)(n-1)/2 - n) = n(n-3)/2$

Note that "f" indicates that the value may be determined by the equation.

> For it is the whole business of humanity in this world to deny evolution, to make absolute distinctions, to take a pen and draw round certain actions a line that nature does not recognise; to take a pencil and draw round the human face a black line that is not there. I repeat, it is the business of the divine human reason to deny that evolutionary appearance whereby all species melt into each other. This is probably what was meant by Adam naming the animals.
>
> GKC, "The Way to the Stars" in *Lunacy and Letters*

How I Came to Invent the Isohodes

As I stated earlier, just a few years ago I began (or resumed) exploration of the Travelling Salesman Problem. In the course of my exploration, I found a novel way of presenting some information about a given graph. As we shall shortly see, this information is defined in the usual formal manner of mathematics. It is not intuitive, magic, or a matter of art, though it certainly may have an artistic quality on occasion. It is readily computable assuming a machine with sufficient speed, and it is easily displayed on a color-graphics device. I named this information the *isohodes* for the given graph, and I will explain that shortly. But first I think it may be both interesting and helpful to tell you how I came to do such an odd thing.

The first chapter of this book sets forth how, back in 1978, before I knew very much about algorithms and complexity, I needed to solve a TSP-like problem for work. I tried to solve it in the usual "brute-force" way, which was quite futile. However, as I also stated, the task at hand was *not* the TSP in its pure form: the desired result was not simply the closed path with shortest Euclidean distance that visits all the given points, nor a Hamiltonian cycle of minimum weight, but a path which would satisfy certain user-directed constraints to find the "best way" through the given coordinates, where "best" is understood from an engineering perspective, and not strictly mathematical in form.

It is this very important distinction which has continually acted as a kind of combination governor-and-exciter through my years of software development: it was always necessary to keep the *user's* purpose[20] and intentions in mind, balancing these with the available data and the demands of the algorithm, its complexity, and the computational machinery available. So powerful is this remarkable distinction[21] that some thirty-odd years later,

---

20 I am well aware that "Purpose" is a very loaded word, coming not from mathematics but from metaphysics, but this is not the place to discuss it, or its preëminent role in computer science. Which probably means another book will have to be written.

21 The Latin word *Distinguo!* (I distinguish) was one of the classic responses of the Scholastics in their formal method of Argument. It was said when a proposed idea must be divided into distinct aspects, each of which are treated independently. (In that era, Argument was not a means of convincing another, but of finding the Truth.) As I note in another context, one naturally learns more from a comparison of two distinct items than from a single one. (See my *A Twenty-first Century Tree of Virtues*, 68.) As Chesterton said, "it is the whole business of humanity in this world to deny evolution, to make absolute distinctions, to take a pen and

I again took up the underlying question to re-examine it in the light of more than three additional decades of experience, graduate-level coursework in algorithms, the reading of innumerable journal articles, attendance at many seminars, talks with colleagues, and the intellectual boost I had from studying the works of G. K. Chesterton, J. H. Newman, S. L. Jaki, and others on questions of the intellect, science, and related topics.

There are some other influences which are worth noting. First is the remarkable advances in computing hardware. I have here at my fingertips a computer (now a few years old) with an address space of 32 bits, that is, a total of 4,294,967,296 bytes, together with a hard-drive space of 1.5 terabytes. The clock speed is roughly one gigahertz, which means it can perform about a billion operations in one second.

The second was the odd little art-form known to some as "fractals" but properly known as the Mandelbrot set: given a point $z = x + iy$ in the complex plane, find the "degree" of the divergence of its square. That is, we count the number of times we may repeatedly compute the square of that complex number until its modulus *diverges* beyond some convenient maximum. This degree is plotted for that coordinate on the complex plane as a pixel of a color determined by that degree. The resulting diagram is not very large; it looks like a kind of gingerbread man on its side, but with a oddly irregular sort of "boundary" between the region of the complex plane which converges and the various regions which diverge. The diagram is even more curious because when one attempts to enlarge the scale for a particular portion of that boundary, very strange things are noted.

(Please note: this is not the place to discuss the matter; I am not getting into something as esoteric as that boundary. This diagram only provided a starting point for a train of thought.)

Many people (even including computer scientists) find the Mandelbrot diagram remarkable because it has a certain "natural" irregular appearance, reminding the viewer of coastlines or other such naturally irregular boundaries. It is quite unnatural since it is an *infinitely* enlargeable irregularity: its ruggedness continues in a quite similar manner no matter how great one magnifies the details of the boundary. This set is also remarkable because the so-called gingerbread man which is its main macroscopic feature may be found countless times in various scales and orientations at ever-finer degrees of magnification.

---

draw round certain actions a line that nature does not recognise..." GKC, "The Way to the Stars" in *Lunacy and Letters*, 78. That title is a citation of Virgil's *Aeneid*, ix, 641, which (as he notes in the essay) is the motto of Edinburgh and is carved over the gate of its castle. At the risk of unduly prolonging this footnote, it is appropriate to cite GKC again: "I would insist that people should have so much simplicity as would enable them to see things suddenly and *to see things as they are.* ... I do not even mind whether they can put two and two together in the mathematical sense; I am content if they can put two and two together in the metaphorical sense." GKC ILN July 13 1907 CW27:506, emphasis added.

Also playing a role in my motivation was the concept of the Voronoi diagram: a graph which is derived from a given (Euclidean) graph which has certain useful properties, and which also reveals deeper facts about the original graph.[22]

Finally, a little tool I built during my work on DNA sequences also played a role. A colleague obtained the complete DNA sequence from the chloroplast (the photosynthetic unit of a plant cell) of the liverwort *Marchantia polymorpha*, and I wrote a very simple program to display all 121,024 bases at once on a color-graphics screen using one pixel per base. After the display was complete, the colors assigned to the various bases could be altered dynamically. It was an exceedingly curious image...

and it suggested among other things that this sequence contained portions which were C-G-rich (as the biologists call it). I did not pursue this exploration, but I kept this trick for handling a huge quantity of data in my toolbox.

None of these are the driving element, however. It was a much simpler idea: the idea that one might iterate a very difficult problem for a given coordinate in order to compute some particular value for that point, and then display its "value" in a kind of three-dimensional picture: that is, using the axis of color to represent a mapping of the third dimension.

* * *

22 But this is getting subtle, so perhaps we should defer further discussion. I hope that's not a Fermat-like cop-out, especially since I can make the margins of *this* book as big as I want! Well, just a little, then, since this is a footnote. The idea of a graph which somehow derives from another graph suggests the concept of "functionals" in calculus: whereas a function is something applied to *numbers* which produces *numbers*, a functional (such as the derivative or integral) is something applied to a *function* which produces a *function*. Such a concept comes from a higher order in the hierarchy of the abstract universe of mathematics, and is well worth pondering: it is a superlative Problem-Solving Skill, and one which cannot be learned if one thinks of calculus as meaning "knowing how to use a website to find an integral." (Oh yeah? But then there are websites that claim to do translation. Ahem!) A competent computer scientist ought to be able to integrate tan(x) by inspection without consulting a table. Possibly there is a deeper reason to learn calculus, just as there is for LONG DIVISION. See the first volume of this series for more on that.

22

The curious fact about this remarkable construct of graphics is that it can communicate a huge amount of data in a very convenient form. It relies on the remarkable properties of the human eye, which is actually a very highly parallel device: some 150 million light-detecting cells in the retina are simultaneously processing the incident light as projected through the lens; this data is processed by a layer of "co-processing cells" (the horizontal, amacrine, and bipolar cells), the output of which is routed through a channel of width only 1 million neurons in the optic nerve. Thus, the eye can "judge" the regions of a given portion of the two-dimensional coordinate system, and this is made easier by using distinct colors.[23]

* * *

But as I said, this diagram was merely a starting point, a clue to a means of presentation. We will now see what I invented.

It was nothing more complex than this. I wanted to understand more about the Travelling Salesman Problem, not only its "universe of discourse," that is, the complete collection of all possible tours which are to be considered, but also the nature of the single tour which is the minimum in that collection. It was easy enough, when given some particular graph, to produce a list of all of its tours – that is, provided there were not "too many" nodes in the given graph.

Now, for someone who has played with lots of different algorithms and riddles and mathematical theorems, it is very clear that for *some* problems, when we have a solution for a given graph, we may be able to use it as a kind of "lever" to derive the corresponding result for a graph with *one additional node*. Hence this sets up a strategy of induction.[24]

So I said to myself: all right, we are given a graph of n nodes, and we have all of its tours, and so we know which one is the minimal length tour. What happens if instead of adding *this particular* point, that is, the "next" one from the graph of $n+1$ nodes, we add *every* point in the plane, using a kind of test-point technique, as in the concept of "neighborhood" in the theory of complex math?

In other words, what if we tried to compute (in the theoretical sense, of course!) a function which mapped every (x,y) coordinate in the entire 2-plane to a tour, the tour which solves the TSP for those $n+1$ nodes? What

---

23   The work of that secondary layer of cells in the retina of the human eye is *edge-detection*. Those cells classify the "kinds" of regions by analyzing the data from their neighbors, so a significant amount of the processing of the raw optical data is performed in the eye, not in the brain. This utterly fascinating topic deserves further study by computer scientists; see any good advanced biology text for more on our wonderful sense of vision.

24   This is "mathematical induction": the principle that if *something* is true for a starting value, and can be shown true for the "successor" of a given value (the value which follows the given one, e.g. is greater by one), then that *something* is true for *all* values in that infinite series. This is a very fundamental concept in the mathematics underlying computer science, and is applied in many ways both in theory (i.e. proofs) and in practice (i.e. algorithms, hence software).

sort of function might that be? Will it have well-defined regions, like the "neighborhood" of a complex function? Or will it have that strange irrational or maybe transcendental boundary, as in the Mandelbrot or other such sets? Just what sort of a function will it be? That is:

What is the "functional character" of such a function (if it even exists)?

Of course I am nowhere as deep into mathematics as I ought to be, and do not have the kind of tools needed to approach the question in the formal theoretical manner. But it was very easy to set up a little test program to see what might happen in an experiment.

And so, after a number of revisions, I contrived a "test lab" which would iterate across a small portion of the xy-plane at a given fixed increment, adding just that single point to a given graph, and performing the exhaustive (and exhausting) computation of every possible tour for the augmented graph, recording the optimal one, and registering it in a list of all possible canonical tours, all the while displaying the corresponding pixel in an appropriate color on the graphics screen.

As you might expect, such a program runs slowly. But it resulted in some amazingly beautiful and rather unexpected diagrams. Let us look at one and talk about it a little.

The Isohodes for a Graph of Four Nodes

We'll start with a small graph, just to see what things look like. We are given the following points, which are the vertices of a regular pentagon of radius 3, but with the lower right node omitted:

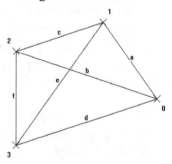

node	x	y
0	3.000	0.000
1	0.927	2.853
2	−2.427	1.763
3	−2.427	−1.763

As it has four nodes, it has six edges:

  a   joins node 0 and node 1
  b   joins node 0 and node 2
  c   joins node 1 and node 2
  d   joins node 0 and node 3
  e   joins node 1 and node 3
  f   joins node 2 and node 3

The addition of a fifth (test) point, node 4, will add another four edges, which are identified by:

  g   joins node 0 and the test point (node 4)
  h   joins node 1 and the test point (node 4)
  i   joins node 2 and the test point (node 4)
  j   joins node 3 and the test point (node 4)

In order to determine the isohodes, we pick a region around this graph, not too large, and some convenient interval, not too small, since for every one of these test points we must solve the Travelling Salesman Problem.[25]

The program is quite easy to arrange. Every computer scientist knows that there is a *very simple* algorithm to solve the TSP, as long as one isn't fussy about how long it takes. All one needs to do is visit all nodes recursively in a depth-first scan of the graph, tracking the total length of the edges traversed in the present tour, and when all nodes have been visited, add in the length of the return edge to the starting node, and see if that total distance is less than that for any other tour. When the recursive nest has completed its factorial-size scan of the graph, we know which tour is the minimum.[26] (See Appendix 3 for the pseudo-code.)

But, as we remarked, on a sufficiently robust machine, even such an algorithm can be performed for reasonably small values, even hundreds or thousands of times. Our usual experimental range was from −5 to +5 in both *x* and *y*, with an increment of 0.02, hence to do one such diagram requires solving the TSP over 250,000 times.[27]

---

25   Granted there are only *five* points in this case, which means we must compute the total distance for just 12 tours, but we will probably want to do other and larger graphs, so we have to be judicious in arranging our experiments.

26   The issue is speed, of course: the simple recursive algorithm requires checking $(n–1)!/2$ tours. As my friend Jim Waclawik has remarked, "the problem isn't the problem; the problem is the solution." Though this remark is a paradox, as we shall see later.

27   Hey, I wasn't doing the work, so I didn't mind. And computers don't sweat. Why should it sit there, flashing its cursor and twiddling its electronic thumbs while I sit here hard at work, thinking up projects for it to do? I like to keep my machinery busy, just like that manager at Reading Sheet Metal.

But for five nodes, there are only 12 tours to be checked, so in a couple of seconds this very interesting image appeared:

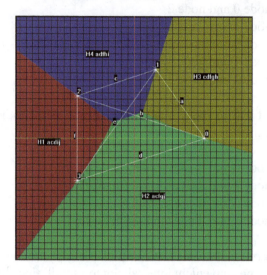

The given graph is shown in white lines, and the four nodes (0, 1, 2, 3) and six edges (a, b, c, d, e, f) are labelled. But there are also four colored regions and which are labelled "H2 acfgj" and so on, and they are the reason for our experiment.

These regions are the *isohodes* for this graph. Each colored pixel represents the (x,y) coordinate for a test point (the unlabelled node 4), and the actual color at that pixel represents one of the four distinct solution tours for the TSP which could occur for the *extended* graph of five points (that is, the given graph with the addition of that particular test point).

I don't have the very first such diagram I generated, but it is irrelevant anyway. The simple fact is that the result was so stunning to me (though I am very easily amused, as you might have guessed) that any one of these diagrams is stunning. Some, as we shall see, are even more surprising, but let us proceed.

So we have performed the exhaustive and exhausting recursive TSP algorithm on some 250,000 distinct cases, tracked all the optimal solutions (in terms of their canonical edge tours) and marked that coordinate accordingly in the appropriate color. What did we learn from this little chore?

Well, before we discuss that, let's try to get a better handle on what we are seeing. Remember what we are doing. We are repeatedly adding a test point, a fifth node to our given graph: a node which would be labelled node 4, and brings in edge "g" to node 0, edge "h" to node 1, edge "i" to node 2, and edge "j" to node 3.

Our diagram reveals that when that point is added anywhere in a particular region as shown by the distinct colors, the solution to the TSP has the same minimal tour as all other coordinates of that same color. You may refer to the above image or this little chart:

where	label	tour	color
left	H1	acdij	red
bottom	H2	acfgj	green
right	H3	cdfgh	yellow
top	H4	adfhi	blue

Consider, for example, the green region at the bottom, labelled "H2 acfgj." Any test point in that region, added as node 4 to the graph, has the minimum-length TSP tour called "acfgj" which is its canonical edge form.

But for the blue region in the upper left, any point added as node 4 has "adfhi" as the solution tour of the TSP.

At first this was surprising to me; I had expected something a bit more complex, maybe even Mandelbrot-like. But then, as I thought about it, I found it was not all that surprising at all. Just as in the concept of the "neighborhood" in complex math, shifting the location of any given node by a "trivial" amount[28] should emphatically *not* alter the solution of the TSP. Hence there really should be "regions" having equal solution tours.

Maybe I was expecting to find *more* regions... I am not sure, but the remarkable utility of having such a curious display for inspiration was a great stimulus to exploration.

What Are These Things?

After playing with my little tool for a while, I began to study the results with some attention to detail... but the very first issue was what were these things to be called?[29] Well, these regions have the property that any point within them gives rise to the *same* minimal-distance tour: that is, they are regions of *equal ways*. In a kind of worn-off sort of Greek we may call them *isohodes*, since ισος (*isos*) means "equal, same as" and 'οδος (*hodos*) means "way, road."

To put it simply, an isohode for a given graph on the Euclidean 2-plane is a region within which the addition of any point in that region to the graph gives rise to the *same* TSP solution tour as others in that region.

So the graph in the above example may be said to have four isohodes, four regions within which any additional point gives rise to the same solution tour.

---

28 Thus suggesting something akin to limit-theory with "some epsilon greater than zero."
29 It may offend certain readers, but I cannot help that, for in writing this I immediately thought of Genesis 2:19 where Adam names the animals. See also GKC, quoted above.

A Formal Definition

Let the tour $t_{TSP} = TSP(G)$ be the solution tour to the Travelling Salesman Problem for any graph G of n nodes in the Euclidean 2-plane.

We define an *isohode* H for a given graph G to be the set of all points r in the plane which have the *same* TSP solution when G is extended by the addition of r:

$$H = \{(x,y)\}$$

such that for any $p, q \in H$,

$$t_{TSP} = TSP(G \cup p) = TSP(G \cup q)$$

For convenience in notation, we define TSP(H) to be the solution t_{TSP} for an isohode H.

Also, for a graph G of n nodes, we define the set of isohodes I(G) to be the set of all isohodes for that graph. By construction, we must have the union over all members of I(G) equal the complete Euclidean 2-plane, since there cannot be a point in the 2-plane which is not contained in any isohode. It may be the case that some points will belong to more than one isohode; as we shall see, such points occur on the boundaries between these regions. (Whether it is possible for an isohode to contain only one point I leave for future research.)

The program to display isohodes for a given graph is quite easy to devise, though it can take quite some time to run depending on the size of the graph, and the granularity and range of interest – but as you have already seen, it produces very interesting results.

* * *

Now, since I am both a computer scientist and a hotshot daredevil programmer with a brand-new toy to play with, I wanted to simultaneously start probing into the theory and continue experimenting with different graphs. For the sake of a tidy presentation, it will make more sense to present a few more examples first, and then begin to examine the theory, even though both paths quickly led me onto even more curious and unexpected paths.

Isohodes for Graphs of Three Nodes

Given that any "interesting" graph of three nodes is always a triangle, and it has one single tour which is always the solution, it should be interesting to see what the isohodes look like for this simplest case. As it turns out, there really are three distinct cases to examine, based on which kind of triangle one has.

An Equilateral Triangle

Here is a nice simple graph, an equilateral triangle of radius 3:

node	x	y
0	3.000	0.000
1	−1.500	2.598
2	−1.500	−2.598

As it has three nodes, it has three edges:

a   joins node 0 and node 1
b   joins node 0 and node 2
c   joins node 1 and node 2

The addition of the fourth (test) point, node 3, adds another three edges:

d   joins node 0 and node 3
e   joins node 1 and node 3
f   joins node 2 and node 3

Here is the diagram of its isohodes:

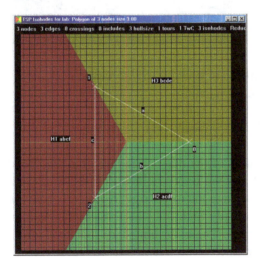

The three isohodes are bounded by the bisectors of the angles of the triangle, and have the following tours:

where	label	tour	color
left	H1	abef	red
bottom	H2	acdf	green
top	H3	bcde	yellow

Isosceles Triangle

Next we have an isosceles triangle:

node	x	y
0	4.000	0.000
1	−1.500	0.750
2	−1.500	−0.750

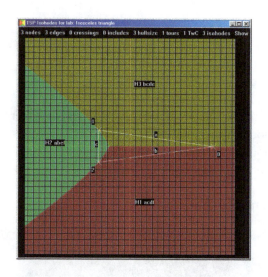

where	label	tour	color
left	H2	abef	green
bottom	H1	acdf	red
top	H3	bcde	yellow

Careful examination of the diagram reveals that the boundary between the left region (H2) and the upper right region (H3), as well as that between the left region (H2) and the lower right region (H1) are *curves*, not straight lines. They are actually limbs of a hyperbola; the equation is discussed in detail in Appendix 1. The other boundary (between upper right and lower right) is a straight line.

Scalene Triangle

And now, a scalene triangle:

node	x	y
0	4.000	0.000
1	−1.500	2.500
2	−2.500	−2.500

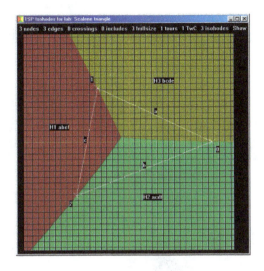

where	label	tour	color
left	H1	abef	red
bottom	H2	acdf	green
top	H3	bcde	yellow

Though it might not be apparent from this diagram, all three boundaries are curved; they are all limbs of hyperbolae.

As you have noted, in each case so far examined there are exactly three isohodes, which partition the plane so as to include all of a particular edge of the original graph. The boundaries always include the given vertices, and they have a conjunction somewhere within the triangle. All this, as well as the actual equations of the boundaries, are readily derived.

Another Scalene Triangle

However, consider this somewhat different scalene triangle, which results in something unexpected:

node	x	y
0	1.000	−0.500
1	−1.000	−0.500
2	3.100	1.500

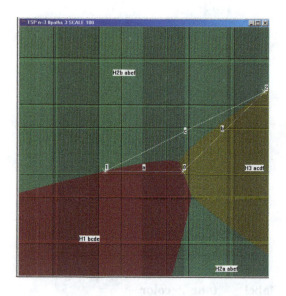

where	label	tour	color
lower left	H1	bcde	red
mid right	H3	acdf	yellow
top, lower right	H2	abef	green

Here we see that there are still three isohodes,[30] but one of them (H2) consists of two "limbs" or sub-regions, H2a and H2b. Now the theory demanded attention: it is clear that there are *only three* possible tours for a graph of four nodes, so there cannot be more than three isohodes, but the fact that one of them had split was annoying. It took some exploration of the mathematics to find out what was going on with those curves, but that was not very difficult; it is merely a case of applying the distance equation for the test point, and so the curves proved to be (portions of) limbs of hyperbolae.

---

30 Just in case you are wondering because node 2 is quite close to the right-hand border of the diagram, I should here point out that I have performed these experiments on a larger range than I have shown in these diagrams. Hence I can assure you that there are no other isohodes lurking beyond the borders.

What is going on? These regions are places in which a test point falls into a certain relation to the given points based on its distance to them. At the boundaries of these regions, the relation degenerates into a relation of one sum: that of a variable distance (between the test point and a fixed point) and a fixed distance, to a similar sum coupling a different fixed point and fixed distance. Such an equation is almost certainly that of a conic section, and the details showing a particular case to be a hyperbola are presented in Appendix 1. For example, the boundary for the red H1 region in the lower left is the limb of the hyperbola which has its axis aligned with edge b, goes through node 1, and has nodes 0 and 2 as foci. Likewise, the boundary of the yellow H3 region on the right has its axis aligned with edge a, goes through node 2, and has nodes 0 and 1 as foci.

As I said, this result nudged at theoretical issues, and raises several questions:

(1) How many isohodes will a graph of n nodes have?
(2) What is the maximum for n nodes, and the minimum?
(3) What controls the splitting of isohodes into two?
(4) May there be more than two such limbs or sub-regions?

But graphs with just three nodes are the simplest case of graphs, and we need more data. Besides, as interesting as this theory is, we are aiming for a better understanding of the TSP, and we don't want to get lost in a forest of alternative and possibly misleading tours... so let us examine some larger graphs.

Graphs of Four Nodes

When there are four nodes, we have two major sub-cases. The first is the square and any of its variants, be they rectangles, rhomboids, trapezoids or even mere quadrilaterals. The important detail is that no single point falls within the triangle formed by the other three, which constitutes the second major sub-case.

First sub-case: quadrilaterals (the regular tetragon)

We start with the regular tetragon, also known as the square.

node	x	y
0	3.000	0.000
1	0.000	3.000
2	−3.000	0.000
3	0.000	−3.000

There are six edges:

a joins node 0 and node 1
b joins node 0 and node 2
c joins node 1 and node 2
d joins node 0 and node 3
e joins node 1 and node 3
f joins node 2 and node 3

The addition of the fifth (test) point, node 4, adds another four edges:

g joins node 0 and node 4
h joins node 1 and node 4
i joins node 2 and node 4
j joins node 3 and node 4

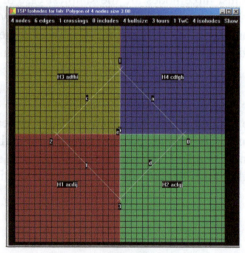

Unfortunately, the labels of the horizontal edge from node 0 to 2 ("b") and the vertical edge from node 1 to 3 ("e") overlap in this diagram. This graph of a regular 4-gon results in *four* isohodes:

where	label	tour	color
lower left	H1	acdij	red
lower right	H2	acfgj	green
upper left	H3	adfhi	yellow
upper right	H4	cdfgh	blue

This seems to suggest that a regular k-gon will have k isohodes – but let's not jump ahead too fast. First we'll see what happens if it's not regular.

A Quadrilateral

We have already examined this one, a 4-gon of unequal angles and sides, but we show it again here, for comparison with the regular 4-gon in the previous example.

node	x	y
0	3.000	0.000
1	0.927	2.853
2	-2.427	1.763
3	-2.427	-1.763

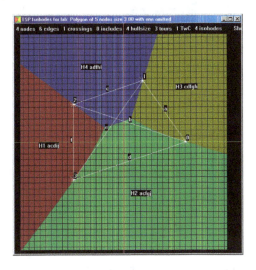

As noted before, we have four isohodes:

where	label	tour	color
left	H1	acdij	red
bottom	H2	acfgj	green
right	H3	cdfgh	yellow
top	H4	adfhi	blue

Note that they are the same four tours as in the square case, though in a different order and with different shapes.

Second sub-case: triangle plus an interior point

The alternative to a quadrilateral graph is a triangle with a single node in its interior.

node	x	y
0	0.000	4.000
1	3.464	−2.000
2	−3.464	−2.000
3	0.000	0.000

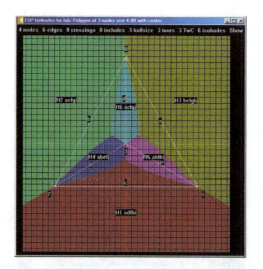

What a nice diagram![31] Now things have gotten interesting. There are four nodes and six isohodes, and they have very curious shapes; the boundaries are definitely curved, or at least some of them are. The first three isohodes are unbounded, partitioning the 2-plane outside of the triangle, though each includes a portion of the interior. The other three isohodes are petal-like wedges aligned with the bisectors of the three angles.

where	label	tour	color
bottom	H1	adfhi	red
upper left	H2	aefgi	green
upper right	H3	befgh	yellow
lower left wedge	H4	abeij	blue
lower right wedge	H5	abfhj	magenta
vertical wedge	H6	acfgj	cyan

This diagram suggests the question: are the portions of isohodes which extend beyond the outermost edges of the graph always in one-to-one relation with those edges, or do any of the "interior" isohodes ever extend beyond

---

31 You gotta love color graphics, especially if (like me) most of your early years in computing were spent with punch cards and line-printer listings, teletypes, paper tapes, or slow monochrome displays. Did you ever stop and think how powerful a tool it is – and then be grateful for the science and engineering that makes it possible?

those edges? That is, can more than one isohode cross a bounding edge of the graph (or, as it is phrased in graph theory, an edge of its convex hull)?

Scalene with an Interior Point

Let us try that again, but with a scalene triangle.

node	x	y
0	−1.000	0.000
1	2.500	0.500
2	0.000	2.000
3	0.750	0.500

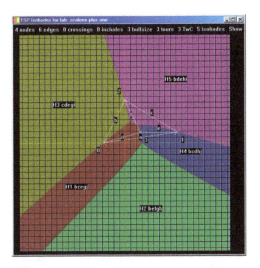

As you can see, this answers some of the questions we just proposed. This graph has five isohodes sharing seven boundaries:

where	label	tour	color
lower left wedge	H1	bcegj	red
bottom	H2	befgh	green
upper left	H3	cdegi	yellow
right wedge	H4	bcdhj	blue
upper right	H5	bdehi	magenta

I looked further into these boundaries, and the details of their functional character are set forth in Appendix 1.

Graphs of five nodes.

After all my years dealing with wacky problems, I had the instinctive feeling that all those graphs (with just three, or four nodes) were "simple" and so I rather distrusted my work. Sure, they were simple things, and they had simple solutions. They were *small* graphs! It wasn't until I began looking at five-node graphs that I felt I would begin to learn something.

Just to review, a graph of five nodes (0, 1, 2, 3, 4) has ten edges, along with five additional ones linking those nodes to the test point:

from\to 0	1	2	3	4	test
0 -	a	b	d	g	k
1	-	c	e	h	l
2		-	f	i	m
3			-	j	n
4				-	o

Of course I started with a nice regular pentagon...

node	x	y
0	3.000	0.000
1	0.927	2.853
2	−2.427	1.763
3	−2.427	−1.763
4	0.927	−2.853

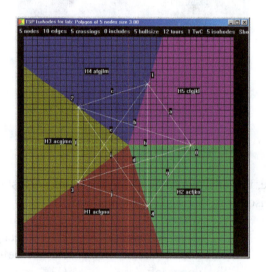

Here are its five isohodes:

where	label	tour	color
lower left	H1	acfgno	red
lower right	H2	acfjko	green
middle left	H3	acgjmn	yellow
upper left	H4	afgjlm	blue
upper right	H5	cfgjkl	magenta

Five nodes, and five isohodes. I shrugged. That felt odd. Well, then, of course it's regular.

A Triangle With Two Interior Nodes

Let's see what happens if I do something irregular, like putting two points inside a triangle...

node	x	y
0	3.000	0.000
1	−1.500	2.598
2	−1.500	−2.598
3	0.999	0.035
4	−0.530	0.848

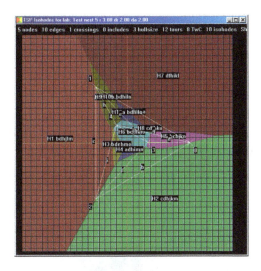

Oh man, that's very interesting! Five nodes, and *ten* isohodes! Even if I could make the diagram larger, it would be difficult to describe all the distinct regions of these isohodes, especially the small ones within the triangle, so I will just give a reduced list:

isohode	tour
H1	bdhjlm
H2	cdhjkm
H3	bdehmo
H4	adhimn
H5	bchjkn
H6	bcdhno
H7	dfhikl
H8	cdfhko
H9	bcdjlo
H10	bdhiln

At least something is starting to make sense, since there are a total of ten edges, and since when we add our test point we must find a tour that breaks at most one existing edge in the graph to insert the pair of edges which link in the test point... Ah, this suggests that any given graph of *n* nodes and $e=n(n-1)/2$ edges cannot have any more than *e* isohodes. Tricky stuff.

We also note that we again have isohodes extending outward from within the bounding edges of the graph (that is, beyond its convex hull).

This sort of thing was more like what I was expecting... so maybe things are starting to make sense... er... well, not really. It's early yet. But... but anyway, I'll keep exploring, since (in those most excellent words of the Great Detective) "the game is afoot."

(No; this particular diagram *was* telling me something, but I could not discern that the message was clearer in the previous instance. Not yet.)

Another Triangle With Two Interior Nodes

Next, let's see what happens when we shift the interior points a little...

node	x	y
0	4.000	0.000
1	−2.000	3.464
2	−2.000	−3.464
3	1.449	0.388
4	−1.061	1.061

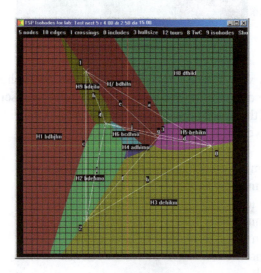

All right, now there are just nine and not ten isohodes; I wondered which one vanished, and why, or if it's just beyond the range being displayed. Again I give a reduced list:

isohode tour

H1	bdhjlm
H2	bdehmo
H3	dehikm

H4 adhimn
H5 behikn
H6 bcdhno
H7 bdhiln
H8 dfhikl
H9 bdeilo

An Irregular Pentagon

How about just an irregular pentagon?

node	x	y
0	3.000	0.000
1	1.000	3.000
2	−2.500	1.500
3	−3.000	−1.500
4	0.500	−2.500

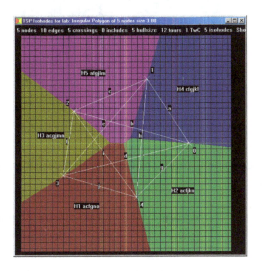

Well... that's like the regular pentagon, just squashed.

(*You're not listening, are you?* it whispers with a nasty giggle. Yeah, it's a graph. Graphs giggle.)

A Square with Central Interior Point

All right, let's make it a square with a single interior point...

node	x	y
0	3.000	0.000
1	0.000	3.000
2	–3.000	0.000
3	0.000	–3.000
4	0.000	0.000

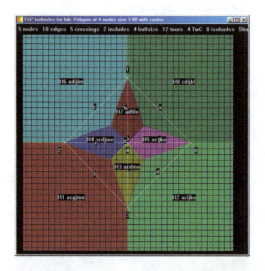

Oh man, that's gorgeous! Eight isohodes with five nodes. I think we can describe this one:

where	label	tour	color
lower left	H1	acgjmn	red
lower right	H2	acijkn	green
lower petal	H3	acdino	yellow
left petal	H4	acdjmo	blue
right petal	H5	acfjko	magenta
upper left	H6	adijlm	cyan
upper petal	H7	adfilo	brown
upper right	H8	cdijkl	dull green

A Square with an Offset Interior Point

What about if that interior node is not right at the center?

node	x	y
0	3.000	0.000
1	0.000	3.000
2	−3.000	0.000
3	0.000	−3.000
4	0.500	0.500

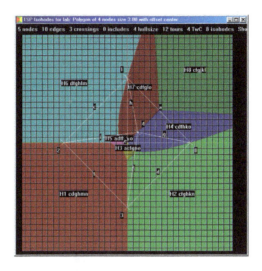

This one also has eight isohodes, though they're hard to read from this particular image. Again we see isohodes (H4 and H7) crossing the bounding edges, that is, those of the convex hull.

Graphs with Six Nodes

Remember, a graph of six nodes (0, 1, 2, 3, 4, 5) has fifteen edges, along with six additional ones linking those nodes to the test point:

from\to 0	1	2	3	4	5	test
0	-	a	b	d	g	k	p
1		-	c	e	h	l	q
2			-	f	i	m	r
3				-	j	n	s
4					-	o	t
5						-	u

43

First, we have the regular hexagon.

node	x	y
0	3.000	0.000
1	1.500	2.598
2	−1.500	2.598
3	−3.000	0.000
4	−1.500	−2.598
5	1.500	−2.598

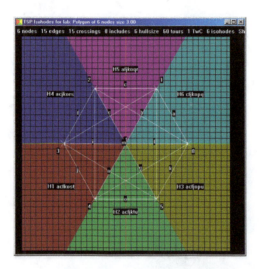

That's just what we expected, given what happened with the pentagon.

where	label	tour	color
lower left	H1	acfkost	red
bottom	H2	acfjktu	green
lower right	H3	acfjopu	yellow
upper left	H4	acjkors	blue
top	H5	afjkoqr	magenta
upper right	H6	cfjkopq	cyan

A Triangle with Three Interior Nodes

Let's go for what looks like the hard one. Let's try a triangle with three interior nodes:

node	x	y
0	4.500	0.000
1	−2.250	3.897
2	−2.250	−3.897
3	2.500	0.000
4	−1.250	2.165
5	−1.250	−2.165

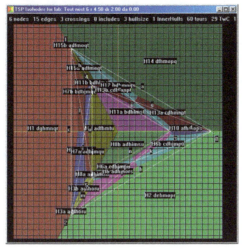

Wow. Just six nodes, a triangle inside a triangle – but 15 isohodes, and even by using various colors it is starting to look like the close-up of a mosaic... I don't think that giving the list of those isohodes will be any help, but here it is:

H1   dghmnqr
H2   dehmopr
H3   adfhoru
H4   adhmnrt
H5   adhmors
H6   cdhjmpu
H7   bdhjmqu
H8   adhimsu
H9   adfhmtu
H10 afhmops
H11 bdhlmst
H12 bdhmoqs
H13 cdhmnpt
H14 dfhmopq
H15 adfmoqt

Just to help you out with this messy diagram, here's the graph without anything else:

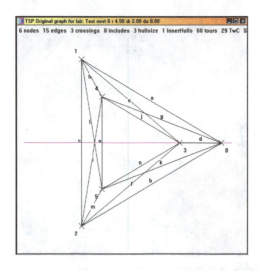

Now, there is something very interesting here. This graph has 6 nodes and 15 edges, but there are *only three* intersections of edges. The graph for the regular hexagon has 15 intersections. Clearly there's something odd going on.

A Pentagon with Interior Node

Let's see what happens when we use the regular pentagon and put one node inside...

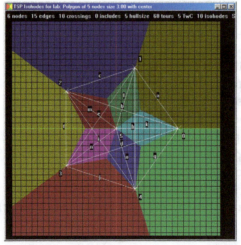

Hm, ten isohodes. But there's still a lot of intersections there.

An Irregular Hexagon

Let's do just one more, an irregular hexagon:

node	x	y
0	3.000	0.000
1	1.000	2.500
2	−2.000	2.000
3	−3.000	0.500
4	−1.500	−3.000
5	2.000	−1.000

Six isohodes... very much like the regular one.

Graphs with Seven Nodes

As the numbers get larger, this little tool runs slower, since it must solve TSP about a quarter of a million times. At $n=7$, the diagram takes nearly a minute to generate. Just for example, the regular heptagon has seven isohodes:

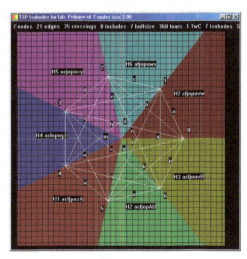

47

And a hexagon with a central point has 12 isohodes:

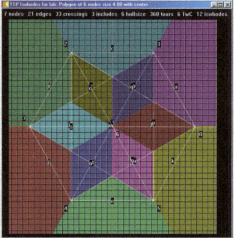

A pentagon with two interior nodes also has 12 isohodes:

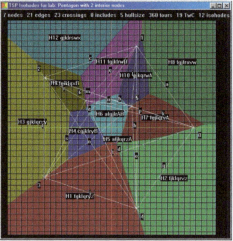

But this triangle (with an interior node) inside a triangle has 14 isohodes:

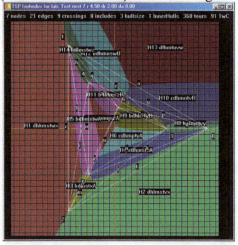

Time To Think?

We could show more fascinating diagrams for more graphs, but perhaps it is time to think about what we are seeing.

The first thing we note is that the regular polygons, and even the irregular ones, always have a very simple arrangement of isohodes. Something about these graphs is *simple*, and adding a new node is somehow "easy" to handle.

There is a dramatic change when we take a polygon and add just one interior node, and yet the situation does not get as intense as adding two, or what is starting to look things get harder when the graph is a triangle with all remaining nodes interior to it. Wait... *nested triangles*...

Then something clicked, and I recalled reading that finding the "depth" of a point in a Euclidean graph of nested triangles constituted a pathological case.[32] The "depth" of a point being considered is the number of successively computed convex hulls which must be stripped from the graph until that point is removed. (This observation about nested triangles proved to be suggestive as we shall see shortly.)

But there were simpler cases... and while it was *easy* enough to understand that there were *hard* cases, it was far *harder* for me to understand why there were also what seemed to be *easy* cases too.[33]

The Number of Isohodes

Before going further, we should make an observation about isohodes which has now become clear. Since a graph with $n+1$ nodes has

$$(((n+1)-1)!)/2 = n!/2$$

possible tours, it can have no more than $n!/2$ possible isohodes. However, after consideration of our results, it is clear that there can be no more than $n(n-1)/2$ isohodes, since this is the number of edges in the original graph, exactly one of which must be replaced with a new pair of edges in order to link the test point into the tour.

---

32 In Preparata and Shamos, *Computational Geometry*, 172-3.

33 Boy, talk about literary chiasms! Er, sorry, a chiasm is a kind of palindromic repetition, a figure of speech which occurs in certain ancient tongues. For example:

And God created man to His own image:

to the image of God He created him... (Gen 1:27)

They are a tell-tale sign of poetry in the Hebrew Scriptures such as the Psalms, and have even been used recently, e.g. in the title *The Origin of Science and the Science of Its Origin*, by S. L. Jaki. A chiasm is merely a literary X-shape, or cross-over, where elements repeat in inverted order: ...αβ...βα... Palindromes (pop, noon, level, Latin *esse*, Spanish *oro*, German *reliefpfeiler*, Greek αιτια) are not merely linguistic oddities: they indicate secondary structure in ribosomal and transfer RNA; see any molecular biology text for details. The term *chiasm* comes from the Greek letter X (*chi*); in anatomy one finds the *optic* chiasm (also called the optic commissure) as portions of the optic nerves cross over, thus forming a *chi* on their way to the brain. See *Gray's Anatomy*, 721.

Some Other Questions About Isohodes

As you will see very shortly, my investigation veered off rather quickly from a merely theoretical study of these very interesting isohodes. But I did spend some time trying to get further with them, at least to the extent that I collected some questions for future exploration.

The first thing I wanted to know was whether such a thing could be computed by some means other than the horribly tedious iteration which solved the TSP thousands and thousands of times in order to draw just one such diagram. Our examples display the region $(-5, -5)$ to $(+5, +5)$ in steps of 0.02, which requires the TSP to be solved about 250,000 times.

It was very funny, and I recall laughing about this when I first coded up my tool and began exploring the isohode diagrams: usually one stands aghast at having to do the TSP *once*, considering every possible tour – and here I was, solving it over and over again, deep within a pair of nested loops! It sounded utterly crazy and foolish.[34] (Was this that risky sort of thing done by hotshot programmers – or was it the dull plodding route taken by scientists trying to boil down a ton of pitchblende?) But of course I kept going, because one never knows what might happen.

1. Short of the exceedingly complex $O(d^2 \cdot (n-1)!/2)$ iteration to gain a d-by-d sample approximation, is there a more direct algorithm to compute the isohodes for a given graph G of n nodes?

2. Can the *number* of isohodes be computed directly from a graph?

3. Can the precise bounds or other mathematical specification of the isohodes for a graph be made in a rigorous manner in the usual manner of mathematics?

There are others which I did not jot down, but having this novel tool, like the Voronoi diagram, is suggestive in many ways.

Unfortunately, I had begun to suspect something, something remarkably curious and tantalizing, and I decided to begin investigating that matter. And if you ask what matter, I will tell you.

Why are the regular polygons so simple? What is going on? Aren't there *always* $(n-1)!/2$ tours through a graph of n nodes, *all of which must be checked*, in order to find the tour with minimal length?

Or is that "accepted doctrine" somehow inaccurate?

If it was inaccurate, that could have a profound impact on the GREAT QUESTION on which countless computer scientists have puzzled:

What is the relation of P versus NP?

And so, as fascinating as these isohodes were, I turned my attention to the question of complete graphs on regular polygons.

It was (as I thought) a well-trodden path I was embarking on. But you should bear in mind that I had in my pack a new tool that was already beeping its little head off about something strange... and so I began to pay very close attention to those "too simple" isohodes.

---

34 After all, I recalled, "...the foolishness of God is wiser than men..." (1 Cor 1:25)

We now come to the sticking-point, or fork in the trail, wherein I stumbled over a very odd fact. But it is better to play it out, not to strut or to grovel, but in order to help others who may be trying to explore things for themselves. At a later date, people will say it's obvious, but it wasn't obvious before, or someone might have pointed it out. And if it's lurking out there in the literature somewhere, it's very good at lurking.

I was puzzled by the curiously regular arrangements of the isohodes for the regular polygons, and even for those which were merely deformed versions of regular polygons. There was something odd going on there; they seemed simple.

Too simple.

I didn't understand why the addition of a single point to a graph of n nodes at the vertices of a regular n-gon always reduced to exactly n isohodes. The tours were all the same otherwise; there wasn't the complications that arose in the cases where some points were interior... There was another sort of regularity, not nearly as obvious, which seemed to be occurring when there was just one interior point.

Moreover, at least in the case of regular polygons, it seemed that there ought to be *more* alternatives. I shouldn't be seeing this very drastic lopping-off ("tree-pruning" as it is called in certain kinds of algorithms) of branches in the vast factorial-tree of possible tours.

All Those Edge Intersections?

The first odd thing I noted was that the regular n-gons are the graphs which have the *most* intersections among their edges. So I decided to examine the business of edge intersections.

Now, the regular n-gon (like any other complete graph of n nodes) has $e = n(n-1)/2$ edges, and since all of these except for the n edges in the polygon itself are internal to the polygon, there are $n(n-1)/2-n = n(n-3)/2$ such edges.

However, every such edge is going to intersect other edges.[35]

---

35 The total number of intersections of the chords of a regular n-gon (with its nodes perturbed slightly in order to avoid duplicate intersections at the center) is given by:

$$\binom{n}{4} = \frac{n(n-1)(n-2)(n-3)}{24} = \frac{n^4 - 6n^3 + 11n^2 - 6n}{24}$$

Just in case you are wondering how to arrive at this, it's straightforward. Choose any four nodes (without replacement). Since this is a regular n-gon, all nodes are on the convex hull, and thus there are no interior nodes, which means those four points form a unique tetragon, the diagonals of which intersect in a unique point. There are $\binom{n}{4}$ ways of choosing those four nodes, hence the above equation specifies the number of intersections.

It is easy to see that all (n–3) interior edges emanating from a given node must intersect the edge which joins its immediate neighbors on the polygon.

But don't other graphs with the same number of nodes have the same number of intersections? Well, in a word, *no*; in fact, we have already seen this in our examples. Consider the case where $n=4$, for which all complete planar graphs have six edges and may be reduced to either a tetragon, or a triangle with one interior node:

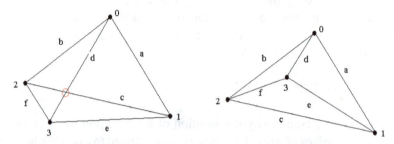

The tetragon has two diagonals which intersect in a single point (in the red circle), but a triangle with an interior point has *no* intersections.

Similarly, graphs with five nodes have ten edges. The regular pentagon has five edge intersections, and a square with one interior point has three, but a triangle with two interior points has only one:

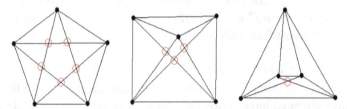

We should here state that there is no complete graph with five nodes which has *no* intersections, since K_5, the complete graph on five nodes, is provably non-planar.

Consider next the case where $n=6$: the 15 edges of a regular hexagon (with its vertices slightly perturbed so the diameters have distinct intersections) has 15 intersections, a pentagon with one interior node has ten, a square with two interior nodes has nine, but a triangle nested within another has only *three* intersections. (Hmmm, nested triangles...)

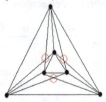

In each case, there are *fewer* isohodes, hence *fewer* tours, in those graphs

52

where there were *more* intersections. Perhaps there was something about these graphs...

Crossings Are Not Allowed

It didn't take very much to discover a simple result from planar graph theory, that optimal paths never had any crossings, and soon I had worked up the proof for what I called the Edge Crossing Theorem:

Edge-Crossing Theorem. (Forbidden Crossings) In a complete graph G of *n* nodes, a cyclic tour $P = \{p_0, p_1, p_2, ... p_{n-1}, p_0\}$ which contains a crossing of edges is *not* a minimal tour; it can always be shortened. (Hence such a tour cannot be the TSP solution.)

Proof. Let the pair of edges that cross be $p_i p_j$ and $p_g p_h$. For simplicity, and since the path is a cycle, we shall denote P by $s_1 p_i p_j s_2 p_g p_h$, where s_1 and s_2 are the remaining nodes in the path. Thus, the total distance of that path is

$$dist(P) = dist(s_1) + dist(p_i p_j) + dist(s_2) + dist(p_g p_h)$$

Since the edge between p_i and p_j crosses the edge between p_g and p_h, we have the following situation:

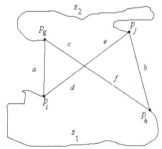

where the length $dist(p_i p_j) = d + e$, and $dist(p_g p_h) = c + f$. Thus, the total length of the path P is

$$dist(P) = dist(s_1) + dist(p_i p_j) + dist(s_2) + dist(p_g p_h)$$
$$= dist(s_1) + (e + d) + dist(s_2) + (c + f).$$

Next, we consider an alternative path $Q = s_1 p_i p_g s_2 p_j p_h$ which avoids the crossing. Its total distance is

$$dist(Q) = dist(s_1) + dist(p_i p_g) + dist(s_2) + dist(p_j p_h)$$
$$= dist(s_1) + a + dist(s_2) + b$$

However, since G is in the Euclidean 2-plane, by the triangle inequality, we have:

$$a < c + d$$

and

$$b < e + f$$

Therefore, we have

$dist(Q) < dist(P)$.

Hence, P is *not* minimal, and thus it cannot be the TSP solution for G. ***Q.E.D.***

Of course this has probably been known since Euclid, and if I had paid more attention in class, maybe I would have recalled it sooner. (There's probably a much tidier proof in the literature, and if you are a grad student, and not too tired, you can go and hunt it up for yourself.) But it is important, and perhaps one of those things which is too large to be noticed,[36] and so it is good to think about it all over again.

We note that this theorem only tells us that certain pairs of edges will not appear in the TSP solution, and that is a great thing, because we can now state apodictically:

We do not *need* to check *every* tour in order to solve the TSP.

Why? Because any path containing edges which intersect each other *cannot* be part of the tour which solves the TSP, since it cannot be minimal. Well, that's great, we now know that the complexity of TSP is formally smaller than $n!$, though it's not clear by how much.

I wasn't quite through with my exploration. There was something curious about that polygon... those "exterior" or "bounding edges" which is known as the Convex Hull of a given graph.

The Convex Hull

The *convex hull* CH(G) for a given graph G of n nodes in the Euclidean 2-plane is defined as the cycle of $m = |CH|$ edges:
$$c_1c_2, \quad c_2c_3, \quad ... \quad c_{m-1}c_m, \quad c_mc_1$$
which may also be specified by the *series* of m nodes:
$$CH(G) = <c_1, c_2, ... c_m>$$
where each of the c_i are nodes in G, such that for every adjacent pair of nodes c_i, c_j in the hull, all other nodes of G are on the *same side* of the edge c_ic_j, that is, the mathematical line containing both c_i and c_j. The convex hull of a general graph in the 2-plane may contain edges not in G, but all graphs we consider are complete, hence all edges of the convex hull of G are in G.

It is easy to compute the convex hull for a given graph: see any algorithms reference for efficient ways, e.g. *Computational Geometry* by Preparata and Shamos. The complexity for the Graham scan given there (pp. 108-9) is $O(n \log n)$.

The convex hull is made up of the edges of G which are its "exterior" edges, the edges of its bounding convex polygon. As you are about to see, the convex hull for a given graph provides the master governing character

---

36 The *locus classicus* of this famous and important concept is Chesterton's "Three Tools of Death" in *The Innocence of Father Brown*: "Perhaps the weapon was too big to be noticed," said the priest.

of... well, be patient, we are almost there. First we will need one more definition, those particular edges which join the nodes of the convex hull, which we will call its *chords*.

Chords of the Convex Hull

We are given a complete graph G of n nodes and its convex hull CH(G); we assume that the convex hull contains more than three edges. Then we define a *chord* of the convex hull to be any edge in G which joins nodes on the hull, but where that edge itself is *not* in the hull.

That is, a chord of the convex hull is any edge in the graph which has its endpoints on the hull (it begins and ends on the hull), but where that edge itself is not a part of the hull.

Here is another way of saying this: the edge $n_p n_q$ is a chord provided that (1) the convex hull has more than three edges, (2) both nodes n_p and n_q are on the hull, and (3) there is at least one node on the hull between n_p and n_q regardless of the direction travelled around the hull.

As I studied the isohodes and the tours being proposed as solutions, I began to notice that the chords of the convex hull never appeared in those tours. In fact, I was slowly led to conclude that such edges are significant in seeking a solution to TSP, because they *can never be in the minimum-length tour*, and may *always* be excluded from consideration. It wasn't hard to devise a theorem to establish this rigorously:

Forbidden Convex Hull Chord Theorem. In a given graph G, any edge which is a chord of the convex hull of G cannot be in the TSP solution.

Proof. Given a cycle containing such a chord, we will show that there exists a cycle of smaller length. Without loss of generality, let us presume the chord to be more or less vertical. We consider two cases:

1. The paths lead away from the chord in *opposite* directions.
2. The paths lead away from the chord in the *same* direction.

Case 1. Consider a graph G where the chord of the convex hull CH(G) is $n_p n_q$ and let n_L, n_R be the nodes on the convex hull which make $n_p n_q$ be a chord, and not just an edge in the hull.

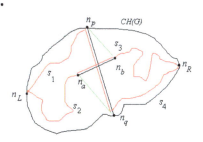

55

Examine the path
$$P_{chord} = n_L s_1 n_p n_q s_4 n_R s_3 n_b n_a s_2 n_L,$$
such that P_{chord} is a Hamiltonian cycle, visiting every node in G. Note that the path approaches the chord from the left, but recedes to the right. Since n_L and n_R are on opposite sides of the chord, there must exist another edge $n_a n_b$, which crosses the chord. It is possible that n_L is n_a, or n_R is n_b, or both in which case the corresponding paths s_1 and/or s_2 will be empty; it is also possible for the other two paths s_3 and/or s_4 to also be empty.

Now, from the previous theorem, we know that there is a shorter path:
$$P_{revised} = n_L s_1 n_p n_b s_3 n_R s_4 n_q n_a s_2 n_L$$
because
$$dist(P_{chord}) = dist(s_1) + dist(n_p n_q) + dist(dist(S_4) + \\ dist(s_3) + dist(n_b n_a) + dist(s_2)$$
and
$$dist(P_{revised}) = dist(s_1) + dist(n_p n_b) + dist(s_3) + dist(s_4) + \\ dist(n_q n_a) + dist(s_2)$$
and (by the triangle inequality)
$$dist(n_p n_b) + dist(n_q n_a) < dist(n_p n_q) + dist(n_b n_a).$$

Hence
$$dist(P_{chord}) > dist(P_{revised}),$$ as was to be shown.

Case 2. Consider a graph G where the chord of the convex hull CH(G) is $n_p n_q$ and let n_L, n_R be the nodes on the convex hull which make $n_p n_q$ be a chord, and not just an edge in the hull.

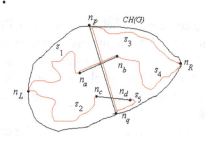

Examine the path
$$P_{chord} = n_L s_1 n_a n_b s_4 n_R s_3 n_p n_q s_5 n_d n_c s_2 n_L,$$
such that P_{chord} is a Hamiltonian cycle, visiting every node in G. Note that the path approaches and recedes from the right side of the chord, but since n_L is on the opposite side of the chord, there must exist two other edges $n_a n_b$ and $n_c n_d$, both of which cross the chord, in order to reach n_L. (It is possible for the three nodes n_a, n_c, and n_L to be one and the same; likewise for n_b, n_d, and n_R.) We may assume that the edge $n_a n_b$ does not cross $n_c n_d$, otherwise we apply Theorem 1, which amounts to merely swapping the left-hand labels of the two edges.)

56

Consider the path

$$P_{revised} = n_L s_1 n_a n_p s_3 n_R s_4 n_b n_d s_5 n_q n_c s_2 n_L.$$

Then we have

$$dist(P_{chord}) = dist(s_1) + dist(n_a n_b) + dist(s_4) + dist(s_3) +$$
$$dist(n_p n_q) + dist(s_5) + dist(n_c n_d) + dist(s_2)$$

and

$$dist(P_{revised}) = dist(s_1) + dist(n_a n_p) + dist(s_3) + dist(s_4) +$$
$$dist(n_b n_d) + dist(s_5) + dist(n_q n_c) + dist(s_2)$$

Thus, comparing these two weights,

$$dist(P_{chord}) :: dist(P_{revised})$$

is the same as

$$dist(n_a n_b) + dist(n_p n_q) + dist(n_c n_d) :: dist(n_a n_p) + dist(n_b n_d) + dist(n_q n_c).$$

Then, adding $dist(n_p n_q)$ to both sides,

$$dist(n_a n_b) + dist(n_p n_q) + dist(n_c n_d) + dist(n_p n_q) ::$$
$$dist(n_a n_p) + dist(n_b n_d) + dist(n_q n_c) + dist(n_p n_q)$$

But from the triangle inequality we know that

$$dist(n_p n_a) + dist(n_b n_q) < dist(n_p n_q) + dist(n_b n_a)$$

and

$$dist(n_b n_d) + dist(n_c n_q) < dist(n_b n_q) + dist(n_c n_d).$$

Hence

$$dist(P_{chord}) > dist(P_{revised}),$$ as was to be shown.
Q.E.D.

It Sounds Nice, Almost Interesting. But...

Now, let us again consider a graph of five nodes which is a pentagon, either regular or somewhat squashed, it makes no difference. It has ten edges, and its convex hull is the pentagon – so by our theorem, none of the other five edges, all chords of the convex hull, may appear in the TSP solution...

But that means this particular case has a trivial solution: the convex hull is the minimal-length tour.

Yeah... that sounds nice... almost interesting. But is it true?

It sure looked that way. I checked the other regular polygons, and some irregular ones, though all of these were convex polygons – and it always seemed to be true. Well... if it seems that way, let's try to state a theorem and see what happens.

The CH=n Theorem. A graph G of *n* nodes for which the convex hull CH(G) contains *every* node of G has the trivial TSP solution of that convex hull.

Proof. This follows directly from the above theorem, since in such a graph all edges except those in the hull are chords, and hence are excluded from

57

the TSP solution.

I remember seeing this somewhere,[37] which I recalled as a rather trite sort of aside... "Well, this is one of those trivial easy cases, like when the matrix is all zeros, so of course there's an easy answer. Duh. Let's get back to whining about how big the factorial grows..." Please note, I mean no disparaging of the eminent scholars who pursue this riddle. At least they had noticed! But I wondered why *I* had never noticed this before, and why nobody had said anything further about it. It seemed like a profound observation, and almost... almost unsettling.

Wait. Let's say it again. That means... for any graph of *n* nodes where the convex hull has *n* edges, there exists a O(1) solution to the TSP.

Did I just type that?

...there exists a O(1) solution to the TSP.

Yeah, I know there was a qualifier to it, a fairly stringent qualifier, an almost absurdly useless-in-the-real-world sort of qualifier. The graph has to be its own convex hull, which sounds rare, and almost contrived. (Bosses don't like rare and contrived things that are useless in the real world.) But my theoretical world just inverted,[38] and I laughed.

That stringent qualifier demands that all the nodes of the given graph are all on the "outside" – that is, all *n* of them are on the convex hull.

But if that was true, I had to ask, what happens in the case when just one point, or a few points, are interior to the convex hull? Is the *n*-edge convex hull case just the odd (meaning "easy") pathological case, like the case of computing the determinant for a matrix which is all zeros, or row-reducing a matrix which is already diagonal?

Or is this oddness just a hint towards other anomalies?

What About Just One Interior Node?

Well, let's see what happens when we are given a graph G of *n* nodes, and there is a single interior point which we'll call n_k. Besides the edg-

---

37 Applegate *et al*, *The Travelling Salesman Problem*, 33; Gutin and Punnen give it as "Lemma 61" with "Folklore" in parenthesis (pp. 548-9).

38 "We were talking about St. Peter," he said; "you remember that he was crucified upside down. I've often fancied his humility was rewarded by seeing in death the beautiful vision of his boyhood. He also saw the landscape as it really is: with the stars like flowers, and the clouds like hills, and all men hanging on the mercy of God." GKC, "The Fantastic Friends" in *The Poet and the Lunatics*.

es of the convex hull, the only edges which can possibly be part of the minimum-length tour must be the $n-1$ edges which join that single interior point to the nodes on the hull; we are never allowed to use any of the hull's chords. But there are only $n-1$ ways of arranging this, since to make the minimal-length tour we must remove one of the edges of the hull and add the edge which runs from one node n_i in the hull to the interior point n_k, and the edge which runs out to the next node n_j in the hull...

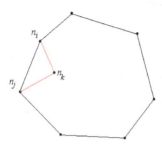

But that means...

...that means we can find the minimal-length tour by a simple loop which tries all $n-1$ possibilities. Which is a $O(n)$ algorithm.

Let's see if this can be established formally...

The CH=n–1 Theorem. A solution to TSP for a graph G of n nodes in which the convex hull contains *all but one* of the nodes of G can be found in linear time in the number of nodes.

Method: Let n_k be the single interior node of G. Assume we are given the convex hull CH(G); we can compute its length $|CH(G)|$ in time $n-1$. We then perform the following test:

For each of the $n-1$ edges $e_{ij} = n_i n_j$ in the convex hull:
    Consider the cycle $n_j...n_i n_k n_j$ produced by eliminating that edge
        and linking the internal node.
    Its length is given by:
$$dist(n_j...n_i n_k n_j) = |CH(G)| - dist(n_i n_j) + dist(n_i n_k) + dist(n_k n_j)$$
The path with the smallest such length is the TSP solution.

Proof. No other paths than these need be considered, as any such path will either (1) contain a chord of the convex hull, or (2) require the crossing of the edges which link the interior node to the others, and hence (by the above theorems) its length will not be minimal. As there are only $n-1$ edges in the convex hull, there are only $n-1$ paths to be tested.

Hence the run-time is $\theta(n)$, not counting the time for the computation of the convex hull. *Q.E.D.*

So, when the convex hull for a graph of n nodes contains $n-1$ edges, there exists a O(n) algorithm to solve the Travelling Salesman Problem.

I wasn't laughing any more; I was dazed. This was really strange. Of course that first case was simple, and it might be an anomaly. And maybe even this case is too... but now we have two data points, and all the inductive machinery is starting to kick into action. Let's see what happens with two.

No... this time we'll go right to the theorem.

The CH=n–2 Theorem. A solution to TSP for a graph G of n nodes in which the convex hull contains *all but two* of the nodes of G can be found in quadratic time in the number of nodes.

Method: Let the two interior nodes of G be designated n_k and n_h. Assume we are given the convex hull CH(G); we can compute its length |CH(G)| in time $n-2$. We perform the following two tests.

Test 1. For each edge $e_{ij} = n_i n_j$ in the convex hull, apply these two steps:

Step 1. Consider the cycle $n_j...n_i n_h n_k n_j$ produced by eliminating that edge and joining in the pair of internal nodes. Its length is given by:
$$dist(n_j...n_i n_h n_k n_j) = |CH(G)| - dist(n_i n_j) + dist(n_i n_h) + \\ dist(n_h n_k) + dist(n_k n_j).$$

Step 2. Consider the cycle $n_j...n_i n_k n_h n_j$ produced by eliminating that edge and joining in the pair of internal nodes. Its length is given by:
$$dist(n_j...n_i n_h n_k n_j) = |CH(G)| - dist(n_i n_j) + dist(n_i n_k) + \\ dist(n_h n_k) + dist(n_h n_j).$$

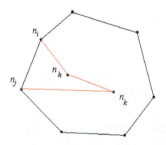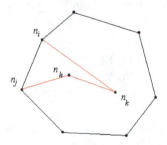

(Note: there will be edges to be tested for which the replacement edges will intersect, but given the simplicity of this method, such a check will gain little.)

Test 2. For each distinct pair of edges $e_{ij} = n_i n_j$ and $e_{pq} = n_p n_q$ in the convex hull, apply these two steps:

Step 1: Consider the cycle $n_i n_h n_j...n_p n_k n_q...n_i$ produced by eliminating edge e_{ij} and linking in n_h, then eliminating e_{pq} and linking in n_k. Its length is given by:

$$dist(n_i n_h n_j...n_p n_k n_q...n_i) = |CH(G)| \quad - dist(n_i n_j) + dist(n_i n_h) + dist(n_h n_j)$$
$$- dist(n_p n_q) + dist(n_p n_k) + dist(n_k n_q)$$

Step 2: Consider the cycle $n_i n_k n_j...n_p n_h n_q...n_i$ produced by eliminating edge e_{ij} and linking in n_k, then eliminating e_{pq} and linking in n_h. Its length is given by:

$$dist(n_i n_k n_j...n_p n_h n_q...n_i) = |CH(G)| \quad - dist(n_i n_j) + dist(n_i n_k) + dist(n_k n_j)$$
$$- dist(n_p n_q) + dist(n_p n_h) + dist(n_h n_q)$$

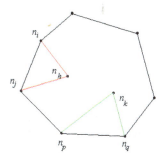 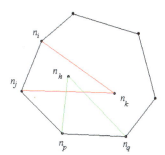

(Note: as mentioned earlier, there are cases where the edges to be tested will cross, and others where they do not. Here we see the arrangement where the new edges will intersect. Again, there is little to be gained by adding such a test to this method at present.)

The path with the smallest length (whether from Test 1 or Test 2) is the TSP solution.

Proof. Any other cycle will contain a chord of the convex hull, and so will introduce a crossing, hence we have enumerated all possible cycles which visit all the nodes of G.

The first test requires inspection of $(n-2)$ tours by two different constant-time computations. The second test requires inspection of a total of $(n-2)(n-3)/2$ tours by two different constant-time computations. Therefore, the algorithm implied by this corollary requires

$$2(n-2) + (n-2)(n-3)$$
$$= (n-2)(n-3+2)$$
$$= (n-2)(n-1)$$
$$= n^2 - 3n + 2$$

Thus the run-time is $\theta(n^2)$, not counting the time for the computation of the convex hull. **Q.E.D.**

The Case where CH=n–3

A solution to TSP for a graph G of n nodes in which the convex hull contains *all but three* of the nodes of G can be found in cubic time in the number of nodes. In this case, we only sketch the method.

There are three separate tests to be made:

Test 1: This test checks the solutions arising by replacing *one* of the edges in the convex hull by one of the six possible paths linking the three interior points.
 Test 1 requires $6(n–3)$ checks.

Test 2: This test checks the solutions arising by replacing *two* of the edges in the convex hull in two steps:
 Step 1: the first edge is replaced by a path linking any one of the three interior points. This requires $3(n–3)$ tests.
 Step 2: the second edge is replaced by a path linking one of the two possible arrangements of the two remaining interior points. This requires $2(n–4)$ tests.
 Test 2 thus requires $3(n–3)·2(n–4) = 6(n–3)(n–4)$ checks.

For example, with n=7, label the edges in the CH a,b,c,d and the interior points 1,2,3.

In step 1, we associate any edge with any point:
 a1, a2, a3, or b1, b2, b3, or c1, c2, c3, or d1, d2, d3

In step 2 we associate the two possible permutations of the remaining points (2,3) or (1,3) or (1,2) with one of the remaining edges:
 a1 and b23 or c23 or d23 or b32 or c32 or d32
 a2 and b13 or c13 or d13 or b31 or c31 or d31
 a3 and b12 or c12 or d12 or b21 or c21 or d21
and so on. Thus there are $12·6 = 72$ possible cases, which is $6(7–3)(7–4)$.

Test 3: This test checks the solutions arising by replacing *three* of the edges in the convex hull by paths linking to each of the three interior points. In order to compute the total number of cases we must consider the $(n–3)(n–4)(n–5)/6$ possible *combinations* of $(n–3)$ edges, pairing each of these combinations with one of the $(3)(2) = 6$ possible *permutations* of the three interior points: (123, 132, 213, 231, 312, 321)

The total number of cases for Test 3 is therefore
 $6·(n–3)(n–4)(n–5)/6 = (n–3)(n–4)(n–5)$.

For example, with $n=7$ and three interior nodes, call the CH edges a,b,c,d, and the interior nodes 1,2,3. There are four possible *combinations* of the four CH edges (abc, abd, acd, bcd), each of which is to be paired with one of the six possible *permutations* of interior points (123, 132, 213, 231, 312, 321). Hence there are 24 possibilities:

    a1b2c3, a1b3c2, a2b1c3, a2b3c1, a3b1c2, a3b2c1
    a1b2d3, a1b3d2, a2b1d3, a2b3d1, a3b1d2, a3b2d1
    a1c2d3, a1c3d2, a2c1d3, a2c3d1, a3c1d2, a3c2d1
    b1c2d3, b1c3d2, b2c1d3, b2c3d1, b3c1d2, b3c2d1

Alternatively, we can derive the number by considering that one particular interior point can be paired with $(n-3)$ different edges. Then there are $(n-4)(n-5)/2$ combinations of the remaining edges, which can be paired in either of two ways, so the total requires $3(n-3)\cdot 2((n-4)\cdot(n-5)/2) = 3(n-3)(n-4)(n-5)$ checks. In our example we again have 24 possibilities:

    a1 together with b2c3, b3c2; b2d3, b3d2; c2d3, c3d2.
    b1 together with a2c3, a3c2; a2d3, a3d2; c2d3, c3d2.
    c1 together with a2b3, a3b2; a2d3, a3d2; b2d3, b3d2.
    d1 together with a2b3, a3b2; a2c3, a3c2, b2c3, b3c2.
for a total of $4\cdot 6 = 24$ possibilities.

So the total checks required by all three tests is
$$6(n-3) + 6(n-3)(n-4) + (n-3)(n-4)(n-5)$$
which expands to
$$n^3 - 6n^2 + 11n - 6$$
or
$$(n-3)(n-2)(n-1)$$

Thus the run-time is $\theta(n^3)$, not counting the time for the computation of the convex hull.

The Case where CH=n–4

A solution to TSP for a graph G of n nodes in which the convex hull contains *all but four* of the nodes of G can be found in quartic time in the number of nodes. In this case, we only sketch the method.

There are four separate tests to be made:

Test 1: This test checks the solutions arising by replacing *one* of the edges in the convex hull by a path linking all four interior points. There is just **one** way of grouping all the interior points which are to be linked.
    We must consider the $(n-4)$ possible *combinations* of choosing *one* CH edge, pairing each of these combinations with one of the 24 possible *permu-*

tations of the four interior points.

Test 1 requires $24(n-4)$ checks.

Test 2: This test checks the solutions arising by replacing *two* of the edges in the convex hull by paths linking to one of the **three** groupings of the interior points into *two* partitions.

We must consider the $(n-3)(n-4)/2$ possible *combinations* of choosing *two* CH edges, pairing each of these combinations with one of the 24 possible *permutations* of the four interior points, either (1) a|bcd, or (2) ab|cd, or (3) abc|d. Hence test 2 requires
$$3 \cdot 24 \cdot (n-3)(n-4)/2 = 36(n-3)(n-4) \text{ checks.}$$

Test 3: This test checks the solutions arising by replacing *three* of the edges in the convex hull by paths linking to one of the **three** groupings of the interior points into *three* partitions.

We must consider the $(n-3)(n-4)(n-5)/6$ possible *combinations* of choosing *three* CH edges, pairing each of these combinations with one of the 24 possible *permutations* of the four interior points, either (1) a|b|cd, or (2) a|bc|d, or (3) ab|c|d. Hence test 3 requires
$$3 \cdot 24 \cdot (n-3)(n-4)(n-5)/6 = 12(n-3)(n-4)(n-5) \text{ checks.}$$

Test 4: This test checks the solutions arising by replacing *four* of the edges in the convex hull by paths linking to each of the interior points. We note that there is only **one** way of grouping the interior points into *four* partitions.

We must consider the $(n-3)(n-4)(n-5)(n-6)/24$ possible *combinations* of choosing *four* CH edges, pairing each of these combinations with one of the 24 possible *permutations* of the four interior points, a|b|c|d. Hence test 4 requires
$$24 \cdot (n-3)(n-4)(n-5)(n-6)/24 = (n-3)(n-4)(n-5)\,(n-6) \text{ checks.}$$

So the total checks required by all four tests is
$$24(n-4) + 36(n-3)(n-4) + 12(n-3)(n-4)(n-5) + (n-3)(n-4)(n-5)(n-6)$$
which expands to
$$n^4 - 10n^3 + 35n^2 - 50n + 24$$
or
$$(n-4)(n-3)(n-2)(n-1)$$

Thus the run-time is $\Theta(n^4)$, not counting the time for the computation of the convex hull.

A Generalization
So far we have the following results:

CH	checks
n	1
$n{-}1$	$(n-1)$
$n{-}2$	$(n-2)(n-1)$
$n{-}3$	$(n-3)(n-2)(n-1)$
$n{-}4$	$(n-4)(n-3)(n-2)(n-1)$

It appears that the number of checks at each level (that is, depending on the number of interior nodes of the graph) is enlarging according to a product of adjacent descending integers (a kind of semi-factorial[39]) which may also be represented as the number of permutations of n things taken k at a time (without replacement):

$$(n)(n-1)(n-2)...(n-k+1) = \prod_{i=n-k+1}^{n} i = \frac{n!}{(n-k)!}$$

We shall now attempt to generalize the equation hinted at by the above table, which will give us the complexity of the case where there are k interior nodes in a graph of n nodes, and $(n–k)$ nodes in the convex hull. In order to do this, we will try to understand what is going on at each test.

Trying To Understand...

It may be a bit confusing to proceed in this manner, but it may help you see how I was trying to explore this. The mathematicians may laugh at what may seem trivial, and yet such details can be very confusing, and even easier to make mistakes with. As I will shortly give a more tidy study, your patience is requested. Moreover, it was helpful to do this detailed study, since it provided a verification on the specific results already reported.

Clearly, since there are k interior nodes, we will need to make k tests, beginning with test (1), in which just one edge is removed from the convex hull, and proceeding until test (k) in which k edges are removed. Regardless of the value of k, at level i, test (i) is the same. Here are the first few, which are easy enough to formulate in a general form:

General Test 1. Choose any of the $(n–k)$ edges of the convex hull. Omit this edge, and link in any covering path through the k interior nodes. There are $k!$ such paths, hence this test must consider $(n–k)\cdot k!$ instances.

Check: as we have seen above for $k{=}3$, test (1) requires $6(n–3)$ checks, which is what this form of the general case specifies.

---

39 One explicit way of defining these "semifactorials" is this:

$$\text{Semifactorial}(p,q) = (p)(p+1)...(q-1)(q) = \prod_{k=p}^{q} k \text{ for } p \le q, \text{ and 1 otherwise.}$$

This avoids confusion about the subscript/superscript notation used in some texts for these upper (or lower) factorials, based on whether the integers ascend or descend.

General Test 2. Choose any two edges (without replacement) of the $(n–k)$ edges in the convex hull. There are $(n–k)$ $(n–k–1)/2$ ways of doing this. Then, letting j vary from 1 to $(k–1)$, perform the following step:

Replace the first edge with a path that visits j of the interior nodes. There are $k!/(k–j)!$ of these paths.

Replace the second edge with a path that visits the remaining interior nodes. There are $(k–j)!$ of these paths.

Hence the total number of instances to be examined at each step is
$$k!/(k–j)! \cdot (k–j)! = k!$$
and there are $(k–1)$ such steps, so for each selected pair we must consider $(k–1)\cdot k!$ instances.

Thus this case must handle a total of $(n–k)\cdot(n–k–1)\cdot(k!)(k–1)/2$ instances.

Check: as we have seen above for $k=3$, test (2) requires $6(n–3)(n–4)$ checks, which is what the general case specifies.

General Test 3. Choose any three edges (without replacement) of the convex hull. There are $(n–k)$ $(n–k–1)(n–k–2)/(3!)$ ways of doing this.

Then, letting i vary from 1 to $(k–2)$, perform the following steps:

(a) Replace the first edge with a path that visits i of the interior nodes. There are $k!/(k–i)!$ of these paths.

(b) Replace the second edge with a path that visits from one up to $k–i$ interior nodes.

(c) Replace the third edge with a path that visits the remaining interior nodes.

At the i-th step there are $(k–i–1)(k!)$ instances to examine, which is summed over all steps to give:
$$(k!)(k–1)(k–2)/2$$
instances for a given trio of edges in the convex hull. Hence the total number of instances considered in this step is
$$((n–k)(n–k–1)(n–k–2)/6)\cdot(k!)\cdot(k–1)(k–2)/2$$
which is
$$(n–k)(n–k–1)(n–k–2)\cdot(k!)(k–1)(k–2)/12.$$

Check: as we have seen above for $k=3$, test (3) requires $(n–3)(n–4)(n–5)$ checks, which is what the general case specifies.

So... what does the general case look like?

i	total for test (i)		
1	$(k!)\cdot(n{-}k)$		
2	$(k!)\cdot(n{-}k)(n{-}k{-}1)$ ·		$(k{-}1)/2$
3	$(k!)\cdot(n{-}k)(n{-}k{-}1)(n{-}k{-}2)\cdot$	$(k{-}1)$	$(k{-}2)/12$
4	$(k!)\cdot(n{-}k)(n{-}k{-}1)(n{-}k{-}2)(n{-}k{-}3)$ ·	$(k{-}1)$ $(k{-}2)$	$(k{-}3)/144$

Well.. this didn't help, or at least nothing came to mind as I worked over these results. Let's try another approach.

* * *

First, we'll summarize our method where the convex hull has $n{-}4$ nodes, that is where $k{=}4$, and see if we can't spot the generalizing trick.

Well, no. I will tell you what happened, since it helps remind us that we ought to be familiar with basic number patterns. As I looked over the above method, I spotted the pattern **one, three, three, one** – which are the terms in row three of Pascal's Triangle. You will see it, and other things, more clearly in the following re-statement of the method.

Our method for a graph with 4 interior nodes, that is, the convex hull has $n{-}4$ edges. We must perform *four* tests:

Test (1). Remove (1) of the $n{-}4$ CH edges and link a permutation of the 4 interior nodes, split into (1) partitions.

There are $\binom{n-4}{1} = (n{-}4)$ possible *combinations* of choosing one of the

$n{-}4$ CH edges, and 4! *permutations* of interior nodes, which may split into *ones* in (1) way: {abcd}, so this term is:

$(n{-}4)\cdot4!\cdot1 = (n{-}4)\cdot24$

Test (2). Remove (2) of the $n{-}4$ CH edges and link a permutation of the 4 interior nodes, split into (2) partitions.

There are $\binom{n-4}{2} = (n{-}4)(n{-}5)/2!$ possible *combinations* of choosing

two of the $n{-}4$ CH edges, and 4! *permutations* of interior nodes, which may split into *twos* in (3) ways: {a|bcd, ab|cd, abc|d}, so this term is:

$(n{-}4)(n{-}5)/2\cdot4!\cdot3 = (n{-}4)\,(n{-}5)\cdot36$

Test (3). Remove (3) of the $n{-}4$ CH edges and link a permutation of the 4 interior nodes, split into (3) partitions.

There are $\binom{n-4}{3} = (n{-}4)(n{-}5)(n{-}6)/3!$ possible *combinations* of choos-

ing three of the $n{-}4$ CH edges, and 4! *permutations* of interior nodes which

may split into *threes* in (3) ways: {a|b|cd, a|bc|d, ab|c|d}, so this term is:
$$(n–4)(n–5)(n–5)/3!\cdot4!\cdot3 = (n–4)(n–5)(n–6)\cdot12$$
Test (4). Remove (4) of the $n–4$ CH edges and link a permutation of the 4 interior nodes, split into (4) partitions.

There are $\binom{n-4}{4} = (n–4)(n–5)(n–6)(n–7)/4!$ possible *combinations* of

choosing four of the $n–4$ CH edges, and 4! *permutations* of interior nodes which may split into *fours* in (1) way: {a|b|c|d}, so this term is
$$(n–4)(n–5)(n–6)(n–7)/4!\cdot4!\cdot1 = (n–4)(n–5)(n–6)(n–7)$$
Observe that the various splittings into permutations are 1, 3, 3, 1, which are the combinations of 3-taken-r-at-a-time, the terms in row 3 of Pascal's Triangle. The generalized form of these tests must therefore be something like this.

A Form of the General Method

When there are k interior nodes, the convex hull has $(n–k)$ nodes, so we must perform k tests, varying i from 1 to k. We note that the CH *must* have at least three nodes,[40] so there may be at most $(n–3)$ interior nodes; that is, we must have $0 \le k \le (n–3)$. The i-th test is as follows:
Test (i). Remove (i) of the $(n–k)$ CH edges and link a permutation of the k interior nodes, split into (i) partitions.

There are $\binom{n-k}{i} = \dfrac{1}{i!}\prod_{j=0}^{i-1} n-k-j$ possible *combinations* of choosing (i)

of the $(n–k)$ CH edges, and $k!$ *permutations* of interior nodes, which may

split into i's in $\binom{k-1}{i-1}$ ways. So test (i) contributes this many tours:

$$Q(n,k,i) = k!\binom{k-1}{i-1}\binom{n-k}{i}$$

As there are k such tests, the total number of tours $T(n,k)$ for a graph of n nodes with $k>0$ nodes interior to the convex hull is their sum:

$$T(n,k) = \sum_{i=1}^{k} Q(n,k,i)$$

For convenience we define the combinatoric term $\binom{a}{b} = 0$ when $a<0$

or $b<0$ or $b>a$, consistent with the appearance of Pascal's triangle. Thus

we may change the index to start at zero, since for $i=0$ we add nothing

---

40   If there are less than three, the graph is colinear, and the TSP has a trivial solution. Proof is left to the reader for homework. (Gosh, how I like saying that.)

$$Q(n,k,0) = k!\binom{k-1}{0-1}\binom{n-k}{0} = k!\binom{k-1}{-1}(1) = k!\cdot 0 \cdot 1 = 0$$

So we may express the total number of tours with the following formula:

$$T(n,k) = \sum_{i=0}^{k} k!\binom{k-1}{i-1}\binom{n-k}{i}$$

where the special case $T(n,0)$ is defined to be 1: that is, when there are no interior points, the convex hull is the single tour which solves TSP.

From the cases worked in detail and the results of our experiments, we hypothesize that this formula is simply the "semifactorial":

$$T(n,k) = (n-1)(n-2)...(n-k)$$

again with the special base case $T(n,0)$ defined to be 1.

Observe that the number of nodes in the convex hull cannot be less than three: there cannot be more than $n-3$ interior nodes to a graph with n nodes. Hence $T(n,k)$ is not defined if k exceeds $n-3$.

Are These Forms Equal?

I worked at reducing the general form for some time, including an application of Stirling Numbers of the First Kind to handle the semifactorial polynomial. (See Appendix 4.) However, for a while I was not able to demonstrate[41] that the proposed general form for $T(n,k)$ (derived from our method) is equivalent to the reduced form, the semifactorial polynomial:

$$T(n,k) = \sum_{i=0}^{k} k!\binom{k-1}{i-1}\binom{n-k}{i} = (n-1)(n-2)...(n-k)$$

Using software I examined all cases of n and k up to $n=15$, and in each case found that the total number of tours generated by our method agrees with the value given by our proposed general equation *and also* with the hypothesized reduced form using the semifactorial: it seemed correct, so it was just a matter of establishing it formally.

I continued to work at the reduction, and eventually found the appropriate equation in one of my reference volumes.

First, we recall that the combinatoric terms, which are known as the binomial coefficients,[42] obey a relation which causes the rows of Pascal's

41   At times like this, we should recall that even Einstein employed a mathematician to help him out: "...about the same time, in the 1910s, Einstein was already employing a personal mathematician..." Jaki, "The Last Century of Science: Progress, Problems, and Prospects" in *The Absolute Beneath the Relative and Other Essays*, 167.

42   I have been waiting a long time for a chance to cite this remarkable line from one of GKC's early essays: "You cannot evade the issue of God; whether you talk about pigs *or the binomial theory*, you are still talking about Him." GKC, *Daily News* Dec. 12 1903, quoted in

Triangle to be identical from left-to-right and right-to-left:

$$\binom{a}{b} = \frac{a!}{(a-b)!\,b!} = \binom{a}{a-b}$$

We have $\binom{a}{b} = 1$ for $b=0$ or $b=a\geq0$, and $\binom{a}{b} = 0$ for $b>a$ or $a<0$ or $b<0$.

Also, by the definition of factorial, we may express our hypothesized "semi-factorial" formula using a combinatoric term:

$$\binom{n-1}{k} = \frac{(n-1)!}{((n-1)-k)!\,k!} = \frac{(n-1)(n-2)...(n-k)}{k!}$$

and so we may write

$$(n-1)(n-2)...(n-k) = \binom{n-1}{k} k!$$

Now we appeal to a standard reference[43] for the following relation, useful for checking the computation of the binomial coefficients:

$$\sum_{i=0}^{t}\binom{r}{i}\binom{s}{t-i} = \binom{r+s}{t}$$

By substituting

$$r = n-k$$
$$s = k-1$$
$$t = k$$

we have

$$\sum_{i=0}^{k}\binom{n-k}{i}\binom{k-1}{k-i} = \binom{(n-k)+(k-1)}{k}$$

$$\sum_{i=0}^{k}\binom{n-k}{i}\binom{k-1}{k-i} = \binom{n-1}{k}$$

But by the above relation from the symmetry of Pascal's Triangle, we have

$$\binom{k-1}{k-i} = \binom{k-1}{(k-1)-(k-i)} = \binom{k-1}{i-1}$$

Maycock, *The Man Who Was Orthodox*. Emphasis added.
43 Abramowitz and Stegun. *Handbook of Mathematical Functions with Formulas, Graphs, and Mathematical Tables. Applied Mathematics Series 55.* p. 822, 24.1.1.II.B, first equation, here written using different variables so as not to clash with those already in use.

Hence the previous equation reduces to

$$\sum_{i=0}^{k}\binom{n-k}{i}\binom{k-1}{i-1}=\binom{n-1}{k}$$

Multiplying both sides by $k!$ we have

$$k!\sum_{i=0}^{k}\binom{n-k}{i}\binom{k-1}{i-1}=\binom{n-1}{k}k!$$

But that gives us what we hoped for, namely:

$$\sum_{i=0}^{k}k!\binom{n-k}{i}\binom{k-1}{i-1}=\binom{n-1}{k}k!=(n-1)(n-2)...(n-k)$$

We have therefore shown that our hypothesized reduced formula correctly expresses the general form of the number of tours as computed by our proposed algorithm.

The Chromatic Polynomial

In attempting to explore the matter from another direction, I found a discussion[44] of the Chromatic Polynomial for a graph, $P(G,x)$ which computes the number of ways of coloring the nodes of the graph G using at most x colors, such that no two nodes joined by a common edge have the same color. Thus a graph of two nodes with a single edge may be colored in $x(x-1)$ ways, and the complete graph on three nodes (which has the form of a triangle) may be colored in $x(x-1)(x-2)$ ways. In fact, for the complete graph of n nodes, the chromatic polynomial has the form of the semifactorial: $x(x-1)(x-2)...(x-n+1)$.

The coefficients of the expanded polynomial may be expressed using the Stirling Numbers of the First Kind.

Careful examination of this, correlated with the idea of considering the paths for the TSP which rely on the segregation between those nodes in the convex hull and those interior to it, resulted in another approach to the matter which gives the same reduced form of the equation.

We are given a graph G of n nodes, with its convex hull of $(n-k)$ nodes and k interior nodes. We consider the coloring of just the k interior nodes using n colors, though with a "first" node on the convex hull being colored first. Then the number of ways of coloring the interior nodes is given by the semifactorial $(x-1)(x-2)...(x-n)$.

Now, if instead of letting those integers assigned to the interior nodes be "colors," we have them indicate the *order* in which those nodes are visited (along with those of the convex hull), we therefore have the equation giving the total number of ways in which those nodes may be visited – and this is in agreement with our hypothesis.

---

44 Roberts, Fred S. *Applied Combinatorics*, 111-122.

A Summary of the Results

As stated earlier, we have written software to experiment with this method, and run it for all cases of n and k up to $n=15$. Here is a table showing the resulting computation of $T(n,k)$, the number of permissible tours in a graph of n nodes with k of them interior to the convex hull. (The convex hull therefore has $n–k$ nodes.) These are the *only* tours which must be checked to find the solution of the TSP, since all others include chords of the convex hull and cannot be a solution.

n\k	0	1	2	3	4	5	6	7	8	9
3	1									
4	1	3								
5	1	4	12							
6	1	5	20	60						
7	1	6	30	120	360					
8	1	7	42	210	840	2520				
9	1	8	56	336	1680	6720	20160			
10	1	9	72	504	3024	15120	60480	181440		
11	1	10	90	720	5040	30240	151200	604800	1814400	
12	1	11	110	990	7920	55440	332640	1663200	6652800	19958400

Note that $T(n,n–3) = (n–1)!/2$,[45] because no tours are excluded when the convex hull has only three nodes and has no chords. However, such graphs with all but three nodes in the interior must contain many pairs of edges which intersect, and therefore can never occur simultaneously in a possible tour, though this method does not address the matter.

In fact, even in the case where there are only two interior nodes, there will be tours to be checked which contain intersections of non-chord edges, and this is easy to demonstrate.

Theorem. In a complete Euclidean graph G of n nodes having at least two nodes interior to its convex hull, there is at least one pair of edges (not chords of the convex hull) which intersect.

Proof. Since there are two (or more) interior nodes, the graph must contain at least five nodes, as the convex hull can contain no less than three. Let Γ be any complete subgraph of G with five nodes, at least three of which (a, b, c) are on the convex hull, and at least one of which (q) is within the triangle bounded by those three nodes. The other interior node (p) may be either inside that triangle or outside it, though it is still within the convex hull of the given graph. By a well-known theorem of graph theory, the complete graph on five nodes, K_5, is non-planar. Hence the subgraph Γ is non-planar, so it contains at least one pair of edges that intersect. We now must show that they are not chords of the convex hull.

---

45   This is readily shown:

$$T(n, n–3) = (n–1)(n–2)...(n–(n–4))\cdot(n–(n–3)) = (n–1)(n–2)...\cdot4\cdot3 = (n–1)!/2$$

which is the number of (canonical) tours for a graph of n nodes.

Observe that the convex hull of the subgraph Γ is *either* (1) a triangle (a,b,c) containing both interior nodes p and q, *or* (2) a quadrilateral (c, a, p, b) containing the other interior node q, which is also within (a, b, c).

In case (1), the only edges which can intersect are those which link hull nodes (a, b, c) to interior nodes (p and q). Such edges are not chords of the convex hull of G. Hence we have established our theorem.

In case (2), let p be the interior node of G which is also a node in the convex hull of Γ, that is (c, a, p, b) where nodes a, b, c are on the convex hull of the original graph. Note that node p extends this triangle into a quadrilateral, the convex hull of Γ. Also consider the edge joining a and b, and note that p and c are on opposite sides of that edge. Finally, let q be the other interior node of G which is also within the triangle (a, b, c).

Four edges link p to the four nodes of Γ. The edges from p to a and p to b may be disregarded. The edge which joins p to q intersects the edge from a to b, but that edge is a chord of the convex hull of G.

However, the edge which joins p to c (recall they are on opposite sides of the edge joining a and b) must intersect both that edge *and also* either the edge from a to q or the edge from b to q, because q is interior to that triangle.

Thus we have shown that Γ cannot be represented in the Euclidean 2-plane without two edges which intersect, since it is a complete graph on five nodes, and also that at least one intersection arises from edges which are not chords of the convex hull of the original graph. *Q.E.D.*

We further note that such pairs of intersecting edges do not (cannot) share any nodes, and from our definition of the count of tours containing a pair of edges, we know that there are $2(n-3)!$ tours eliminated from consideration by any such pair. The number of tours eliminated will of course diminish as further pairs are added, since some tours contain more than one such pair. Recall that a graph of n nodes contains $(n-1)!/2$ tours:
$$(n-1)!/2 = (n-1)(n-2) \cdot (n-3)!/2$$
So the fraction of tours eliminated by one such intersecting pair is

$$\frac{2(n-3)!}{(n-1)(n-2)(n-3)!/2} = \frac{4}{(n-1)(n-2)}$$

which diminishes rapidly as n increases, yet there will certainly be some saving of computational effort, especially when dealing with the pathological case where the convex hull is a triangle.

What does this mean in terms of algorithms?

It means that we should be able to "prune" the tree of tours being considered if we add a test to check whether the next edge being proposed for addition to the tour intersects with an edge already chosen. Such a test requires a preliminary (one-time) computation which is $O(e^2)$ or *quadratic* in the number of edges, hence $O(n^4)$ or *quartic* in the number of nodes.

An new algorithm for TSP

We now present an algorithm which computes the solution for the TSP by checking only tours which exclude chords of the convex hull. First we sketch the method of generating the tours to be tested; none of these tours will contain chords of the convex hull, and by our hypothesis the complexity is of order n^k where k is the number of interior nodes (within the convex hull) of a graph of n nodes. Afterwards we shall examine a way of improving the algorithm by considering other edge intersections.

We are given a graph G of n nodes, its convex hull H of $(n-k)$, and the set I of its k interior nodes. (H and I can be found in time $n \cdot \log(n)$ by readily available algorithms.)

The method is very simple. The tour is simply a traversal of the convex hull along its edges, but where at any given hull node, the next edge may be skipped, and a detour made to visit one of the possible permutations of a partition of the interior nodes. The manner of arranging this partition/permutation is not very difficult, but requires attention to detail.

Here is an outline of the method:

Take an arbitrary node on H as the starting point, then visit the nodes on H in successive order, that is, in the order of the convex hull.

To visit node h of the hull we perform all of the following actions:

Let j be the number of remaining (unvisited) interior nodes.
Repeat the following, decreasing j afterwards until it is zero:
>   Generate a permutation of any j remaining interior nodes
>   Then visit the next node on the hull.

This method may readily be handled by a pair of mutually recursive routines: one to visit the hull nodes, the other to generate permutations. The pseudocode is given in Appendix 3.

I also have a working implementation, and will now show the results for the first few values of n. Note that for ease of recognition, I use letters from the *start* of the alphabet to indicate nodes in the convex hull, and those from the *end* of the alphabet for interior nodes.

Remember, these tours (which are possible solutions for the TSP) may contain only edges from three classes:
>   (1) those joining nodes on the convex hull,
>   (2) those joining interior nodes,
>   (3) those joining a node on the convex hull with an interior node.

No chords of the convex hull may appear in these tours.

Possible Tours for *n*=4:

```
Paths for Graph n=4 Hull=4 Interior=0 nodes: abcd
    1: abcd
n  4 k  0 has 1 total paths

Paths for Graph n=4 Hull=3 Interior=1 nodes: abc[Z]
    1: aZbc
    2: abZc
    3: abcZ
n  4 k  1 has 3 total paths
```

Possible Tours for *n*=5:

```
Paths for Graph n=5 Hull=5 Interior=0 nodes: abcde
    1: abcde
n  5 k  0 has 1 total paths

Paths for Graph n=5 Hull=4 Interior=1 nodes: abcd[Z]
    1: aZbcd
    2: abZcd
    3: abcZd
    4: abcdZ
n  5 k  1 has 4 total paths

Paths for Graph n=5 Hull=3 Interior=2 nodes: abc[YZ]
    1: aYZbc
    2: aZYbc
    3: aYbZc
    4: aYbcZ
    5: aZbYc
    6: aZbcY
    7: abYZc
    8: abZYc
    9: abYcZ
   10: abZcY
   11: abcYZ
   12: abcZY
n  5 k  2 has 12 total paths
```

Possible Tours for *n*=6:

```
Paths for Graph n=6 Hull=6 Interior=0 nodes: abcdef
    1: abcdef
n  6 k  0 has 1 total paths

Paths for Graph n=6 Hull=5 Interior=1 nodes: abcde[Z]
    1: aZbcde
    2: abZcde
    3: abcZde
```

```
        4: abcdZe
        5: abcdeZ
n   6 k   1 has 5 total paths

Paths for Graph n=6 Hull=4 Interior=2 nodes: abcd[YZ]
        1: aYZbcd
        2: aZYbcd
        3: aYbZcd
        4: aYbcZd
        5: aYbcdZ
        6: aZbYcd
        7: aZbcYd
        8: aZbcdY
        9: abYZcd
       10: abZYcd
       11: abYcZd
       12: abYcdZ
       13: abZcYd
       14: abZcdY
       15: abcYZd
       16: abcZYd
       17: abcYdZ
       18: abcZdY
       19: abcdYZ
       20: abcdZY
n   6 k   2 has 20 total paths

Paths for Graph n=6 Hull=3 Interior=3 nodes: abc[XYZ]
        1: aXYZbc
        2: aXZYbc
        3: aYXZbc
        4: aYZXbc
        5: aZXYbc
        6: aZYXbc
        7: aXYbZc
        8: aXYbcZ
        9: aXZbYc
       10: aXZbcY
       11: aYXbZc
       12: aYXbcZ
       13: aYZbXc
       14: aYZbcX
       15: aZXbYc
       16: aZXbcY
       17: aZYbXc
       18: aZYbcX
       19: aXbYZc
       20: aXbZYc
       21: aXbYcZ
       22: aXbZcY
       23: aXbcYZ
       24: aXbcZY
       25: aYbXZc
       26: aYbZXc
```

```
27: aYbXcZ
28: aYbZcX
29: aYbcXZ
30: aYbcZX
31: aZbXYc
32: aZbYXc
33: aZbXcY
34: aZbYcX
35: aZbcXY
36: aZbcYX
37: abXYZc
38: abXZYc
39: abYXZc
40: abYZXc
41: abZXYc
42: abZYXc
43: abXYcZ
44: abXZcY
45: abYXcZ
46: abYZcX
47: abZXcY
48: abZYcX
49: abXcYZ
50: abXcZY
51: abYcXZ
52: abYcZX
53: abZcXY
54: abZcYX
55: abcXYZ
56: abcXZY
57: abcYXZ
58: abcYZX
59: abcZXY
60: abcZYX
n  6 k  3 has  60 total paths
```

But Is It Inductive?

As I hinted in the foreword for this volume, I was hesitant about saying anything about any of this. It seemed too... too strange: both too obvious and also too easy to have made mistakes on. And yet the proofs seemed right, and the experiments bore out the implications, and it hung together: there was just the surprise that no one seems to have noticed it before. Of course it was exciting to contemplate what it could mean... even if the hard case when the convex hull is a triangle requires checking of $(n-1)!/2$ tours, and so ends up being hard in the NP sense.

And then, as usual, the inductive machinery kicked in and I asked myself, is there some way of applying this convex hull idea to *that* case in a recursive manner? I set up experiments and puzzled, like the Grinch, until my puzzler was sore, but I couldn't find a way of taking advantage of the inner hull, because the "Forbidden Convex Hull Chord" Theorem does not apply to it. For example, consider the following graph of eight nodes:

node	x	y
0	5.000	0.000
1	-4.000	4.000
2	-3.800	-4.000
3	0.500	1.500
4	-0.500	1.500
5	1.500	1.300
3	-0.400	1.000
4	0.750	1.000

This graph has 28 edges, and just for reference it has 15 isohodes. Its convex hull has three nodes, and the five interior nodes are themselves an "inner" convex hull – it is a triangle containing a squashed pentagon – and it looks like this:

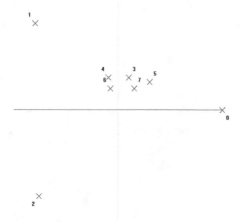

However, as you can see, the solution tour uses two of the chords of the "inner" convex hull:

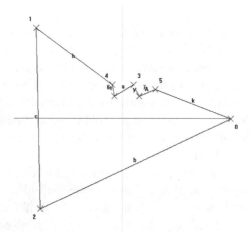

Note that a graph of 8 nodes and 28 edges has 7!/2 = 2520 possible tours. For comparison, here is a little table showing the results of some experiments on eight-node graphs. It gives the number of tours without crossings (twoc) and isohodes (ih):

twoc	ih	graph
1	8	regular octagon
7	14	heptagon with one interior node
27	12	hexagon with two interior nodes
70	17	pentagon with three interior nodes
109	16	tetragon with four interior nodes
212	15	triangle with nested pentagon
339	18	two nested triangles with two nodes inside inner one

Here also is the same table (without isohodes) for graphs of nine nodes, which has a maximum of $8!/2 = 20160$ tours to be checked:

twoc	graph
1	regular nonagon
8	octagon with one interior node
36	heptagon with two interior nodes
99	hexagon with three interior nodes
256	pentagon with four interior nodes
461	tetragon with nested pentagon
729	triangle with nested hexagon
1188	three nested triangles

While we're at it, here's the same thing for ten, which has 181440 tours to be checked.

twoc	graph
1	regular decagon
9	nonagon with one interior node
47	octagon with two interior nodes
156	heptagon with three interior nodes
396	hexagon with four interior nodes
947	pentagon with nested pentagon
1758	tetragon with nested hexagon
1859	triangle with nested heptagon
4080	three nested triangles and an internal node

Huh, an interesting progression. Do these things still feel like they're not growing quite as fast as one might expect? Let's check the ratios:

	total	tri/(n–3)gon			nested triangles		
n	tours	twoc	ratio	cross	twoc	ratio	cross
7	360	67	0.1861	15	91	0.2528	9
8	2520	212	0.0814	32	339	0.1345	30
9	20160	729	0.0362	66	1188	0.0589	36
10	181440	1859	0.0102	116	4080	0.0225	66
11	1814400	6067	0.0033	198	14838	0.0082	110
12	19958400	17525	0.0009	324	75820	0.0038	159

For graphs with n nodes, "tours" is the total possible tours. The next columns are for test graphs consisting of a triangle and an inner polygon of $(n–3)$ nodes: "twoc" heads the column of tours-without-crossings, "ratio" is that number divided by the total tours, and "crossings" gives the number of edge intersections. The last columns are for test graphs of nested triangles with the remaining point(s) within the innermost one: again giving the number of tours without crossings, the ratio of this to the total tours, and the number of edge intersections.

The Proverbial Monkey Wrench

Lest the apparent pattern be a little too dazzling, I should mention that I looked into some other graphs of 11 nodes. I found this curious graph: a triangle around another triangle, around a pentagon. It had more edge intersections than the graph of three nested triangles around two nodes, but it also had more tours-without-crossings! What on earth! Instead of exploring that case further, I found *another* graph of 11 nodes: a triangle around another triangle, around a square, around one inner node, and this one was even worse, or better. Here are some surprising results:

3 triangles, 2 nodes	110 intersections	14838 twoc
2 triangles, pentagon	124 intersections	16350 twoc
2 triangles, square, node	126 intersections	16921 twoc

In fact, there seem to be two kinds of pathological case,[46] depending on what one wishes to examine:

1. Those graphs with the fewest intersections between edges, which are a series of nested triangles of diminishing size, with either zero, one, or two nodes within the innermost triangle. These do not always have the maximal tours-without-crossings, however.

2. Those graphs with the most tours-without-crossings, which follow a pattern depending on n, the number of nodes where $n=3m+r$ and when $r = 1$ we have $(m–1)$ nested triangles, and an innermost square, or when $r = 2$ we have $(m–1)$ nested triangles, an inner square, an innermost node. And yet these do not always have the minimal edge-crossings.

---

46 Tested by experiment for $5 < n < 18$. However, $n=7$ is anomalous: a graph of two nested triangles and an inner node gives 9 edge-intersections and 91 tours-without-crossings, while a graph of a triangle around an inner square gives 15 crossings but only 67 tours-without-crossings.

Even for these difficult cases, the thing which still glares out is that the tours-without-crossings are definitely a good deal fewer than the total number of tours – and we know from the theorems that no optimal tour may contain any edges which cross. We already found one powerful trick which reduces the work dramatically; are there others to be found?

Computing Is, After All, an Experimental Science.

I should stress that all these are *experimental results*, and not formal (algebraic) ones. And yet, clearly something is pruning that factorial-tree. Very curious.

At the same time that little "governor-and-exciter" which I obtained from my experience at FEL is saying very clearly: "Yes, but remember: this is a different riddle now! It is not the same as the original problem, and you cannot expect the same rules to apply. You will have to look into this some more."

Granted. But we do know this: in a complete Euclidean graph, no path which visits all nodes *and crosses itself* can be minimal. Hence, we may *always* disregard any path which contains intersecting edges. Indeed, we may discard from further consideration *any* path of at least three edges which intersects itself. That means the true complexity of the TSP depends on the total number of tours without crossings for the given graph.

Taking the simple case of regular polygons,[47] the number of crossings of edges grows in the following manner:

n	crossings
4	1
5	5
6	15
7	35
8	70
9	126

which (as we have seen) has the recurrence formula:

$$\binom{n}{4} = \frac{n(n-1)(n-2)(n-3)}{24} = (n^4 - 6n^3 + 11n^2 - 6n)/24.$$

Though we know such graphs have trivial TSP solutions, they also have the maximal number of edge-intersections, so this formula provides the upper bound.

What about the lower bound? I was not able to derive a theoretical formula for general graphs, or for the "pathological" cases, so I performed a number of experiments on nested triangles (or with an inner square and node). For each graph, here are the number of edge-intersections (X), the to-

---

47 Note that this assumes a slight displacement of the nodes, so that no three edges intersect in a single point.

tal possible tours (tours), and the number of tours without crossings (twoc).[48]

n	X	tours	twoc	maxentries
5	1	12	8	18 (4)
6	3	60	29	82 (5)
7	9	360	91	368 (6)
8	30	2520	227	1108 (7)
9	36	20160	1188	8267 (8)
10	66	181440	4320	32467 (9)
11	110	1814400	16921	162576 (9)
12	159	19958400	75820	1037808 (10)
13	253	239500800	226082	4371960 (11)
14	349	3113510400	1202902	26272682 (12)
15	483	43589145600	2721060	93366326 (13)
16	666	653837184000	13901436	592419264 (14)
17	994	10461394944000	49616796	2556514152 (14)
18	1282	177843714048000	165761770	11663336845 (15)
19	1690	3201186852864000	692566657	58842803009 (16)

In case you are wondering, the test for n=19 took 17 hours. Well, that column of tours-without-crossings looks very much like alternating powers of two, that is, the powers of four – and plotting it on a logarithmic scale resulted in a very nice almost perfectly straight line, as you can see:

---

48   This table includes anomalous cases (such as n=8) where the graph with minimum edge-crossings is not also the graph with maximal tours-without-crossings. These proposed "pathological" cases (the nested triangles described earlier) are not necessarily the "worst" graphs, though they clearly have fewer intersections than others of the same node count. Hence such cases provide interesting challenges and are definitely not "easy." After all, any graph with a triangular convex hull gains no improvement from our observations about the chords of the convex hull, and if (as it seems) the actual complexity of TSP depends on the number of edge-intersections of the given graph, those graphs with the fewest intersections will be hardest to solve. Presumably the parallel – to cite a radically different problem – would be to find the prime factors of a number like 12894274092511638633349 as opposed to finding them for 13104566989963265480064.

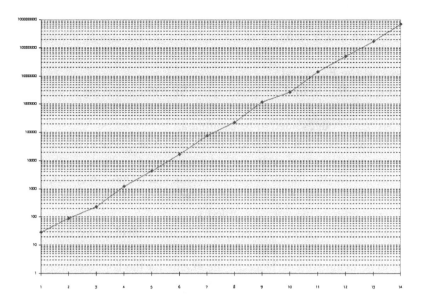

Which fits roughly to something like 4^n, or possibly somewhat less.

Caution is advised against getting one's expectations too high, since the "maxentries" column specifies the maximum number of entries of at some level (in parenthesis) of inner recursion. This happens a few levels before the final one, showing that a good deal of tree-pruning occurs at what might be called "twig" level.[49] This maximum also appears to grow by a power rule, estimated as $y=4.2^x$. At the same time, there clearly is a significant improvement over the "check all tours" method, and this raises a number of questions.

1. For complete Euclidean graphs of n nodes, what is the range of edge-intersections? Is there a general formula for determining how many edge crossings a given graph has? (That for regular n-gons is given above.)
2. What class of graphs have the minimum intersections[50] – is it, as the evidence suggests, those which are nests of triangles?
3. As n increases, how fast does the number of intersections grow? If it indeed grows at 4^n (or less), then pruning by edge-crossing will certainly beat the terrifying growth of that cheerful factorial tree. So it appears, based on the data so far at hand, but this is not just appearance.

---

49   I told you I am rarely serious. I still recall my advisor for my MS talking about "vines" in a "jungle" of semantic trees when he presented a fascinating talk about a representation for formal languages in a paper by Čulik. Our field provides so many more opportunities for humor, and we ought not resist them. Also, see the first Case Studies volume for the story of how I fell out of a tree, and what it had to do with Frankel Engineering.
50   We already know which graphs have the *most*: those where all nodes are in the convex hull. That's how we get that amazing O(1) case for the TSP.

Remember:
No path which visits all nodes and crosses itself can be minimal.

Cross Out the Factorial

It is possible to find *the* solution to TSP without having to examine *every one* of the $(n-1)!/2$ possible tours, since only those tours which contain *no* crossings of edges need to be examined.

Such an algorithm is easy to arrange, The first (given in the appendix as the "more sophisticated recursive solution") simply checks each proposed edge against the previously determined path, and if these intersect, that edge is discarded. (Hence the entire subtree which would have sprouted from adding that edge is likewise discarded.)

The above experiments, up to $n=15$, were checked by both methods (that is, by checking all tours and by pruning those with crossings) and found to be in agreement. Here are the results for the last few cases with the times for "alltours" and "pruned" in seconds:

n	total tours	twoc	alltours	pruned
12	19958400	75820	5.40	.45
13	239599800	226082	44.24	2.17
14	3113510400	1202902	557.17	15.26
15	43589145600	2721060	7964.77	58.78
16	653837184000	13901436	(33 hours)	415.20
17	10461394944000	49616796	(22 days)	2243.65
18	177843714048000	165761770	(1 year)	10964.41
19	3201186852864000	692566657	(18 years)	61270.31

Certainly it's nicer to wait 15 seconds than 10 minutes, and even better to wait *one* minute than to wait 7,964 seconds – which is almost two and a quarter *hours*! I did not run the all-tours routine for $n>15$, but gave estimates, since the run time advances as $n!$. Hence it would take $15 \cdot (2.2$ hours) or over 33 hours to check all tours for $n=16$, compared to the pruning routine which gets the answer in under seven minutes. Yes, I was willing to wait 17 hours, but not 18 years, for $n=19$.

We have found improvements by pruning by other rules, such as discarding from further consideration any partial paths which exceed the minimum solution found so far.[51] This gave promising results as well as revealing unexpected anomalies – but the exploration is continuing, and so we defer further reports to a future edition of this text.

---

51 McHugh's *Algorithmic Graph Theory*, 51 proposes a straightforward method of getting *a* solution to TSP by computing the minimum spanning tree, then transforming that tree into a Hamiltonian cycle; in some cases it may actually find the minimal solution. This simple and efficient preliminary step provides a starting limit to the main algorithm. No tour which solves the TSP – indeed, no path proposed as part of a solution – may exceed the length of that initial tour.

More About Those Edge Crossings

I was annoyed by something about the pathological cases and looked a little further into what was going on. Even in the case of two nested triangles where $n=6$ and there are $(6-1)!/2=60$ possible tours, there are just 29 tours without crossings, even though there are *no* chords of the convex hull. I spent some time examining the list of tours, and soon found that I had not looked quite far enough. I was able to devise a stronger result than the "Forbidden Convex Hull Chord" Theorem:

Corollary (Forbidden Convex Hull Chord-paths).

In a given complete planar graph G of n nodes with convex hull CH, let there be a path $P = <n_g, ... n_h>$ of at least two nodes, of which only n_g and n_h are on the convex hull, and which partitions the nodes of the graph such that there is at least one node on both the "left" and the "right" sides of that path. Then P *cannot* be in any tour which solves the TSP. (The proof follows the same argument as the main theorem and is omitted.)

We can see the result of this corollary from the case where $n=5$.

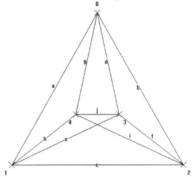

There are two chord-paths:

     nodes 0,4,2 with edges g and i

     nodes 0,3,1 with edges d and e.

Of the 12 possible paths for a five-node graph, four contain edge crossings and of these, three contain these pairs of edges:

     013240   aefgi

     024130   bdehi

     031240   cdegi

This last contains both of the forbidden chord-paths. However, there is one other path with edge crossings which does *not* contain either pair:

     013420   abeij

Clearly, the inclusion of any such chord-path into a possible tour will force an intersection eventually, which will exclude that tour from further consideration. However, excluding such chord-paths, at least in the simple case of *pairs* of edges, may prune the factorial-tree faster than merely checking for edge-crossings.

Note that any number of edges may form such forbidden chord-paths, but it is hard to say whether the complexity of distinguishing these will result in any savings in the larger problem. For example take the case of two nested triangles, which has six nodes, 15 edges, and 60 possible tours, of which 29 have no edge crossings:

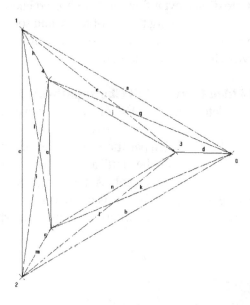

Only 16 of the 31 tours *with* crossings are eliminated by the three chord-paths of *two* edges, g/i (042), k/l (051), e/f (132):

g/i	k/l	e/f
aegimn	beijkl	aefgmo
afgiln	bfhjkl	aefiko
cdgiln	cdijkl	befglo
cegikn	cfgjkl	befhko
degilm	dfhikl	efghkm
efgikl	(efgikl)	(efgikl)

Only six additional tours are eliminated by chord-paths of *three* edges:

l/n/d:	bdhiln
h/o/k:	cdfhko
i/j/d:	adijlm
m/o/g:	cdegmo
e/n/m:	beghmn
h/j/f:	afhjkm

The seven remaining tours with edge crossings contain no path which can be considered a chord of the convex hull. Still, the matter of such forbidden paths is curious, and deserves further study.

Yes, there are lots of paths remaining to be explored.

And that is where I will leave the matter for the present.

One Further Comment

For the moment, it looks like the hard cases are when most of the nodes are interior to the convex hull. The $(n-1)!/2$ possible tours can be pruned down to something on the order of 4^n by discarding those containing edge intersections. Though 4^n is not polynomial in n, and is less than $n!$ for n less than eight, it grows much slower than the factorial:

$$20! = 2,432,902,008,176,640,000$$

but

$$4^{20} = 1,099,511,627,776$$

which is some 2.2 million times smaller. From our experiments, a graph of 18 nodes comprising six nested triangles has $17!/2 \approx 1.77\mathrm{e}14$ tours to consider, but only $1.66\mathrm{e}8$ do not contain intersections.

Granted, these pathological graphs are very unusual – but real-world problems will be likely to have *many* nodes interior to their convex hulls. Nevertheless (as Gandalf said) we cannot guarantee what sort of weather future generations will have. We have to do what we can *now*, in the time we have available, and finding out that *sometimes* there are graphs for which there are deterministic polynomial algorithms to solve the Travelling Salesman Problem may be helpful. In the end, we must consider the purpose, and prudently balance science and engineering in our work.

CHAPTER 5: THE END OF OUR PRESENT PATH

Beatus vir cujus auxilium est a te, Deus,
cum sacra itinera in animo habet.[52]
Blessed is the man whose assistance is from Thee, O God,
when he has the sacred way [road, route, journey] in his soul.

It appears that the GREAT QUESTION has been answered, at least in a partial manner, that great question which has nagged at half a century of mathematicians and computer scientists: "does P equal NP, or doesn't it?"

The answer seems to be *Sometimes*.

Sometimes P = NP.
Sometimes P ≠ NP.

From what we have seen in this little study, it appears that we have been asking the wrong question. We had not really explored the situation, the nature of the problem we were trying to solve. TSP was not a monolithic problem; it contained a wide variety of distinct classes of "problems" ranging from quite easy to very difficult.

This was *not* a matter of not grasping what the customer wanted, as it was with the punch press optimization problem at Frankel Engineering where "optimal" meant "how I think it looks best" – hardly a computable function regardless of considerations of complexity. It is a matter of not realizing that the Travelling Salesman Problem could be segregated into various distinct cases, each of which had its own proper complexity. It was not the problem itself; the Travelling Salesman Problem is neither P nor NP, but sometimes is one and sometimes is the other. Its complexity is inherent in the details of the problem, and a simple procedure applied to the given data provides the classification of the complexity for that case.

Sometimes it takes a little more work to find out what is going on.

Do I think we're done?
No.
Will I, and lots of other computer scientists keep working at this?
Sure.

But then *somebody has to do the hard jobs*.

---

52 From the traditional prayers of preparation before Holy Mass; see *Missale Romanum*, xcv, "*Psalmus 83*" cf. Ps 84:6.

APPENDIX 1: ABOUT ISOHODE BOUNDARIES

We here present two brief studies of the boundaries of isohodes.

First Example: a three-node graph

We are given a scalene triangle, with these coordinates:

 node 0 $(1, -.5)$
 node 1 $(-1, -.5)$
 node 2 $(3.1, 1.5)$

Its three edges are named according to our convention:

 a   joins node 0 and node 1
 b   joins node 0 and node 2
 c   joins node 1 and node 2

The addition of the fourth (test) point adds another three edges, which are identified by

 d   joins node 0 and the test point (3)
 e   joins node 1 and the test point (3)
 f   joins node 2 and the test point (3)

These edges appear in the tours which denote the isohodes of the graph.

Here is the graph along with a representation of its isohodes:

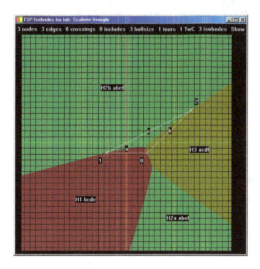

The three isohodes and their tours are:

 H1   bcde   (lower left)
 H2   abef   (upper left and lower right
 H3   acdf   (middle right)

We wish to find the formal expression of these curved boundaries.

The following table presents our results for the three boundaries. They occur at the adjacency of the specified regions (*regions*), at which the two alternative tours (*tours*) are compared. These comparisons algebraically reduce to a single variant term (*variant*) which consists of two variable edge lengths (two of the lengths of edges d, e, or f), and a fixed constant depending on the particular boundary being considered. As we shall show in detail, each of these boundaries are limbs of hyperbolae, or rather portions of limbs.

regions	tours	variant	solution
H1:H2	bcde::abef	d–f+a–c	hyperbola foci 0 & 2 thru 1
H1:H3	bcde::acdf	e–f+a–b	hyperbola foci 1 & 2 thru 0
H2:H3	abef::acdf	e–d+c–b	hyperbola foci 1 & 0 thru 2

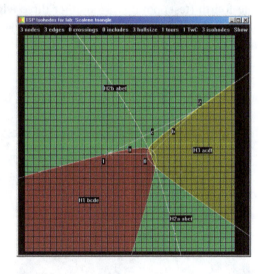

Detailed analysis of boundary H1:H2

Let us consider the case where our test node 3 (x,y) is somewhere to the left of node 1, in the area of the boundary between the lower left region labelled H1, and the upper left region labelled H2b. The set of edges *abc* which is the trivial solution for three nodes, is extended to two possible solutions, either by

    (1) in the case where the test node is in H1, by dropping edge *a*
        and adding edges *d* and *e*, which yields *bcde*,

or

    (2) in the case where the test node is in H2b, by dropping edge *c*
        and adding edges *e* and *f*, which yields *abef*.

The decision as to which solution is minimal rests upon the comparison of the weights for the two edge collections H1 = *bcde* and H2 = *abef*: that is, the comparison of the two weights for the tours <0,2,1,3,0> or <0,1,3,2,0>, the weights of which are:

$$dist(H1) = dist(0,2) + dist(2,1) + dist(1,3) + dist(3,0)$$

with

$$dist(H2) = dist(0,1) + dist(1,3) + dist(3,2) + dist(2,0)$$

The boundary between these two regions is simply the set of points where these two distances are equal:

$$dist(H1) = dist(H2)$$

that is, where

$$dist(0,2)+dist(2,1)+dist(1,3)+dist(3,0) =$$
$$dist(0,1)+dist(1,3)+dist(3,2)+dist(2,0)$$

Recalling that $dist(p,q) = dist(q,p)$, and that these distances or edge lengths are fixed for the given graph, except for those measuring lengths to the test node p_3, we can rewrite this equation as between cd and af, which is

$$dist(1,2) + dist(0,3) = dist(0,1) + dist(2,3).$$

Moreover, the edges c and a, and their weights $dist(1,2)$ and $dist(0,1)$ are fixed by the given graph; only the edges d and f, and their weights $dist(0,3)$ and $dist(2,3)$ are dependent upon the test node p_3, so we can consider the boundary as being specified by the equation:

$$dist(1,2) + dist(0,3) = dist(0,1) + dist(2,3)$$

Expanding these for the arbitrary position (x,y) of the test point p_3, and using the edge names as shorthand for their weights, we have:

$$\sqrt{(x-x_0)^2 + (y-y_0)^2} + c = \sqrt{(x-x_2)^2 + (y-y_2)^2} + a$$

This equation specifies the points forming the boundary between the lower left and upper left regions, that is, the points for which the two tours $bcde$ and $abef$ have the same total distance

Applying algebra, this equation can be revised to the following form:

$$Ax^2 + Bxy + Cy^2 + Dx + Ey + F = 0$$

where the coefficients are given by:

$$A = (x_0 - x_2)^2 - (a - c)^2$$
$$B = 2(x_0 - x_2)(y_0 - y_2)$$
$$C = (y_0 - y_2)^2 - (a - c)^2$$
$$D = 2(x_0\varepsilon + x_2\delta - (x_0 + x_2)\chi)$$
$$E = 2(y_0\varepsilon + y_2\delta - (y_0 + y_2)\chi)$$
$$F = \chi^2 - \delta\varepsilon$$
$$\chi = (\delta + \varepsilon - (a - c)^2)/2$$
$$\delta = x_0^2 + y_0^2$$
$$\varepsilon = x_2^2 + y_2^2$$

(Recall that the values of a and c are simply the distances between nodes 0 and 1, and nodes 1 and 2, respectively.)

This second-degree equation is the hyperbola which goes through node 1 has its foci at node 0 and node 2, and has its axis aligned with edge b, the edge joining node 0 and node 2.

Note that in the special case where a equals c, the hyperbola reduces to a line perpendicular to the line joining nodes 0 and 2.

Second Example: a four-node graph

The coordinates of the nodes are:

0	$(-1, 0)$
1	$(2.5, 0.5)$
2	$(0, 2)$
3	$(0.75, 0.5)$

Its six edges are named according to our convention:

a	joins 0 and 1
b	joins 0 and 2
c	joins 1 and 2
d	joins 0 and 3
e	joins 1 and 3
f	joins 2 and 3

The addition of the fifth (test) point adds another four edges, which are identified by

g	joins 0 and the test point
h	joins 1 and the test point
i	joins 2 and the test point
j	joins 3 and the test point

These edges will appear in the tours which denote the isohodes of the graph.

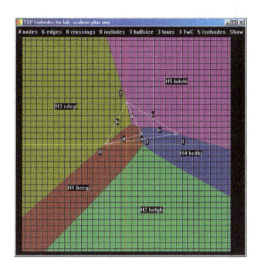

The five isohodes and their tours are:

    H1   bcegj   (narrow diagonal area in lower left)
    H2   befgh   (bottom triangular area)
    H3   cdegi   (upper left)
    H4   bcdhj   (narrow diagonal area on right)
    H5   bdehi   (upper right)

If one studies the regions in detail, that is, if one investigates the functional character of the seven boundaries, one finds something interesting...

regions	tours	variant	solution
H1:H2	bcegj::befgh	j–h +(f–c)	hyp, foci 3 & 1 thru 2
H1:H3	bcegj::cdegi	j–i +(d–b)	hyp, foci 3 & 2 thru 0
H1:H4	bcegj::bcdhj	g–h +(d–e)	hyp, foci 0 & 1 thru 3
H1:H5	bcegj::bdehi	g+j–h–i +(d–c)	(see below)
H2:H4	befgh::bcdhj	g–j +(c+d–e–f)	hyp, foci 0 & 3 constant 1.308452
H3:H5	cdegi::bdehi	g–h +(b–c)	hyp, foci 0 & 1 thru 2
H4:H5	bcdhj::bdehi	j–i +(e–c)	hyp, foci 3 & 2 thru 1

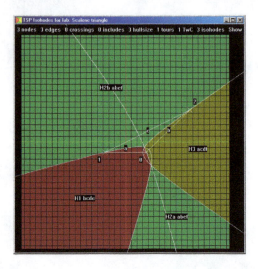

The boundary between H2 and H4 is interesting because it does not go through one of the given nodes, but requires a slightly more complicated constant ($c+d-e-f$) to be computed.

However, the boundary between H1 and H5 is even more interesting because it is not the limb of a hyperbola, and we shall now examine it in detail. I have given it the name "bi-hyperbola" since it seems to be similar to a hyperbola, but with *four* focus-like points, rather than just two.

Consider the boundary between isohodes H1 and H5, in terms of edges, where bcegj = bdehi.

Here, only b and e may be cancelled, which leaves:

$g+j-h-i = d-c$

or, in terms of variables,

$$\sqrt{(x-x_0)^2+(y-y_0)^2} + \sqrt{(x-x_3)^2+(y-y_3)^2} - \sqrt{(x-x_1)^2+(y-y_1)^2} - \sqrt{(x-x_2)^2+(y-y_2)^2} = (d-c)$$

After making little progress in reducing this rather difficult equation, I applied Newton's Method to get some numerical results. This gives an odd-looking curve, similar to a hyperbola, but with a kink in it:

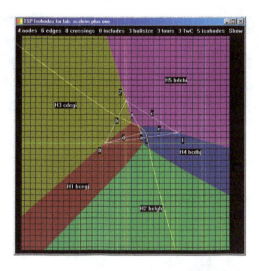

I call it a "bi-hyperbola" since it has a algebraic form similar to that of standard hyperbolae which are defined in terms of a pair of foci and a constant, but this equation has *two* pairs of foci. In this example, those pairs are nodes 0 and 1, and nodes 2 and 3 and the associated constant is determined by edges d and c.

Another way of considering this equation is as the locus of points of intersection of two ellipses, where one has foci at nodes 0 and 3, and the other at nodes 1 and 2. Each of the ellipses is specified in the usual manner as the locus of points such that the sum of the distances from each point to the foci is constant (the value of which is twice its semi-major axis). Moreover, the constant for the first ellipse differs from that for the second ellipse by $(d{-}c)$, the constant in the above equation.

Define the "nest" of ellipses with foci at f_1 and f_2, for any semi-major axis $a_1 > 0$. For each such ellipse, associate with it an ellipse in the nest of ellipses with foci at f_3 and f_4, but with semi-major axis $a_2 = a_1 + (d{-}c)$, with d and c fixed constants as specified above. (Note that we presume the segments joining the foci do not intersect.) Then there is some smallest value a_t at which the pair of ellipses are tangent, having just one point in common, and as we proceed through larger values of a greater that a_t, we shall add further points to the set of intersections. That set of intersections of the two nests of ellipses constitutes the curve for the above equation, as shown in the following diagram.

95

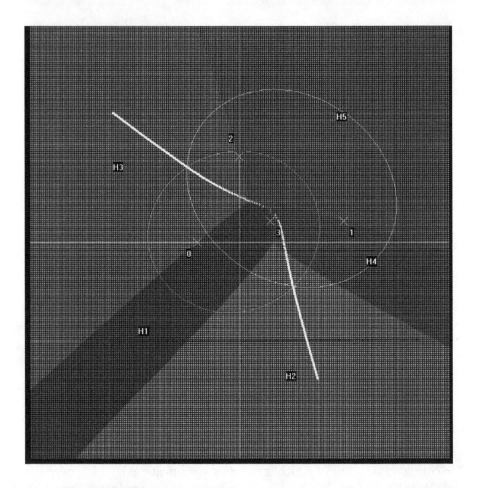

Presumably this is a kind of "hyper-conic" where the curve derives from associations of distances between related points in conics, just as conic equations derive from associations of distances between related points in the plane.

It would be interesting to explore this difficult equation further, as well as the presumed family to which it belongs, but not just now.

APPENDIX 2: THE SHEARING PROBLEM

This problem is closely related to the punch press optimization problem, and I include it as a brief case study.

Sometime in 1978 the engineers at Reading Sheet Metal asked us to look into another problem. They purchased stock sheet metal in rather large sizes, but these were not directly used in the punch press. These stock sheets were first run thorough their shearing machine, which chopped them down to sizes required by the current inventory of parts to be made. Hence, they asked us whether we could give them a computer program which would take as input (1) the number of stock blanks of stated size and (2) a list of currently required parts with size and desired quantity, and produce a series of instructions to chop down the stock into those parts, such that the wasted metal was a minimum.

Rather than discuss this shearing problem in detail, I will here simply state that the theoretical problem is a form of the knapsack problem, and has its own particular complexity. Like the "optimization" problem it presents a challenge... but I devised a program to try to get answers. It is not fair (if I do say so myself) to misjudge my implementation by the standards of a complexity theoretician; indeed, the simple fact was that I actually attained a much more sophisticated result, in that it was multi-threaded. The user started the solution task running, and was able to tell it how long he wished to spend in seeking a solution; he could also query its "best so far" collection and see how good were the results it had found.

However, as in the "optimization" problem, the user showed that he was not really interested in "the best" possible solution, but in one which fitted *his own* concept of "best." It wasn't a computational riddle; it was, perhaps, at best, a form of secretarial duty: he just wanted something to reduce his simple, "best" way of shearing into a sheet of instructions to go to the operator of the shearing machine.

APPENDIX 3: PSEUDO-CODE

This appendix presents pseudo-code for three algorithms:
1. A simple recursive TSP solution which runs in $O(n!)$ time. For a test case[53] with $n=15$ it takes 2.2 hours to examine 4.3e10 tours.
2. A more sophisticated solution which is significantly faster. For the same test case it takes only 58.8 seconds to examine the 2.7e6 tours which are not otherwise forbidden. My estimate (based solely on experimental results) for its complexity is $O(4.2^n)$ with $O(3.6^n)$ tour checks.
3. An algorithm to produce the possible TSP tours for a graph with n nodes, k of which are internal, and the remaining $n-k$ are in the convex hull. It runs in complexity $O(n^k)$.

The Simple Recursive Solution

We are given the graph G of n nodes. We require the following global data structures:

```
int Gn   ;; the number of nodes in the graph
real distance[Gn,Gn] ;; distances between nodes in the graph

doublelist nodelist ;; circular list containing indices to nodes
int path[Gn]   ;; the path presently being considered

real leastdistance ;; the least distance so far found
    ;; it is initialized to "positive infinity" or any convenient
    ;; value larger than the sum of all distances in the graph.
int leastpath[Gn]   ;; the path with the least distance so far found

routine CheckThisPath
arguments
    in float thisdistance
begin
    if thisdistance < leastdistance then
        leastdistance ← thisdistance
        leastpath ← path
    endif
endproc
```

---

53   This is a graph of five nested triangles of diminishing size, shifted so as to have no three nodes co-linear. It has 14!/2 = 43589145600 possible tours, of which only 2721060 contain *no* edge intersections.

```
routine SimpleTSPVisit
arguments
    in integer depth
    in doublelist YetToDo
    in integer prevnode
    in float prevdistance
locals
    doublelist NewList
    listentry thisnode
begin

    if depth = Gn then
        ;; we've done all the nodes
        if path[1]<path[Gn-1] then
            ;; this is a canonical tour, check it
            CheckThisPath(prevdistance+distance[prevnode,0])
        endif
    else
        ;; we have not done all the nodes yet
        NewList ← YetToDo
        for thisnode in NewList
            hold ← thisnode
            RemoveList(thisnode,NewList)
            path[depth] ← hold→this
            SimpleTSPVisit(depth+1,NewList,hold→this,
                prevdistance+distance[prevnode,hold→this])
            AddList(thisnode,NewList)
        endfor

    endif
endproc
```

Calling sequence for SimpleTSPVisit:

```
    ;; first, set the distance array to the
    ;; Euclidean distances between nodes

    ;; also initialize these working variables
    leastdistance ← INFINITY
    path[0] ← 0
    path[Gn] ← 0

    NodeList ← NULL
    for i ← 0 to Gn
        AddList(i,NodeList)
    endfor

    ;; then examine all possible tours...

    ;; CAUTION this is O(n!) and may take a very long time:

    SimpleTSPVisit(1,nodelist,0,0)

    ;; here, leastpath is the path with leastdistance.
```

A More Sophisticated Recursive Solution

This algorithm is derived from the previous one, except that we perform pruning based on whether a proposed edge is forbidden because it intersects with one already in the path.

We are given the graph G of n nodes and $e=n(n-1)/2$ edges. Besides the $n \times n$ matrix of distances between nodes which takes $O(n^2)$ time to construct, we require an array of lists indicating the edges intersecting a given edge; this takes $O(n^4)$ time to initialize. We also require the following global data structures:

```
int Gn   ;; the number of nodes in the graph
int Ge   ;; the number of edges in the graph

real distance[Gn,Gn] ;; distances between nodes in the graph
list edgelist[Ge] ;; which edges this edge intersects

doublelist nodelist ;; circular list containing indices to nodes
int path[Gn]   ;; the path presently being considered
int forbid[Ge] ;; whether or not this edge is presently forbidden

real leastdistance ;; the least distance so far found
     ;; it is initialized to "positive infinity" or any convenient
     ;; value larger than the sum of all distances in the graph.
int leastpath[Gn]   ;; the path with the least distance so far found

routine EdgeIndex
arguments
    in integer i,j
returns
    integer
begin
    if j < i then
        return (i*i-i)/2+j+1
    else
        return (j*j-j)/2+i+1
    endif
endproc

routine CheckThisPath
arguments
    in float thisdistance
begin
    if thisdistance < leastdistance then
        leastdistance ← thisdistance
        leastpath ← path
    endif
endproc

routine SophisticatedTSPVisit
arguments
    in integer depth
    in doublelist YetToDo
    in integer prevnode
    in float prevdistance
```

100

```
locals
    doublelist NewList
    listentry thisnode
    int eindex
    listentry thisedge
begin
    if depth = Gn then
        ;; we've done all the nodes
        if forbid[EdgeIndex(0,prevnode)]>0 then
            ;; cannot use this edge to return home
        else if path[1]<path[Gn-1] then
            ;; exclude non-canonical tours
            CheckThisPath(prevdistance+distance[prevnode,0])
        endif
    else
        ;; we have not done all the nodes yet
        NewList ← YetToDo
        for thisnode in NewList
                hold ← thisnode
                RemoveList(thisnode,NewList)

                eindex ← EdgeIndex(prevnode,hold→this)

                if edgelist[eindex] = NULL then
                    ;; this edge intersects NO other edge
                    ;; add it and go on
                    path[depth] ← hold→this
                    SophisticatedTSPVisit(depth+1,NewList,hold→this,
                        prevdistance+distance[prevnode,hold→this])
                else
                    ;; this edge intersects others...
                    if forbid[eindex] > 0 then
                        ;; do nothing, it's forbidden
                    else
                        ;; it's free, so we can use it

                        ;; first, forbid all edges it intersects
                    for thisedge in edgelist[eindex]
                        forbid[thisedge→this] ← forbid[thisedge→this]+1
                    endfor

                        ;; now add it and go on
                        path[depth] ← hold→this
                        SophisticatedTSPVisit(depth+1,NewList,hold→this,
                            prevdistance+distance[prevnode,hold→this])

                        ;; unforbid the edges
                        for thisedge in edgelist[eindex]
                            forbid[thisedge→this] ← forbid[thisedge→this]-1
                        endfor
                    endif

                ;; all done, proceed to next node

                AddList(thisnode,NewList)
        endfor
    endif
endproc
```

Calling sequence for Sophisticated TSP Visit:

```
;; set the distance array to the Euclidean distances between nodes
;; build the edgecross array of lists of intersecting edges

;; also initialize these working variables

for i ← 0 to Gn
   forbid[i] ← 0
endfor

leastdistance ← INFINITY
path[0] ← 0
path[Gn] ← 0

NodeList ← NULL
for i ← 0 to Gn
   AddList(i,NodeList)
endfor

;; then examine all possible tours...

;; CAUTION this may take a very long time:

SophisticatedTSPVisit(1,nodelist,0,0)

;; here, leastpath is the path with leastdistance.
```

Algorithm for Tours excluding Chords of the Convex Hull

Here is the pseudo-code for the method described in the text for generating tours for a given graph. This code is derived from an actually implemented and running program which was tested through *n*=15. It is a straightforward pair of mutually recursive routines, and uses a handful of global variables defined below. It also requires the prior determination of the convex hull. Note that the following routines are *general*, generating a series of tours indicated by the indices of nodes which are partitioned between those on the convex hull and those interior to that hull. The exact position of the nodes in the Euclidean 2-plane is not needed here.

Global variables

The following global variables are required:

```
integer Gn   ;; the number of nodes in the graph
integer Gh   ;; the number of nodes in that graph's convex hull
integer Gk   ;; the number of interior nodes (Gk = Gn-Gh)

boolean unused[Gn]     ;; scoring matrix for generating permutations
integer thispath[Gn]   ;; the tour to be considered
```

There is also an unspecified routine CheckThisPath() to be called whenever the tour in the global "thispath" is to be checked. Note that the final edge is implied; only Gn entries are made in the array, not Gn+1. (The sample results given in the text were produced by CheckThisPath simply printing the contents of "thispath" at each call.)

Use of the Visit routine

The following initialization must be performed: Gn is set to the node count of the given graph. All Gn members of unused[] are set to TRUE.

Gk and Gh are set based on the convex hull, and each node is assigned its index in the following manner:

When $0 \leq i < $ Gh:
    i is a node *on* the convex hull.
    These nodes must be specified in adjacent order around the hull.

When Gh $\leq$ i < Gn:
    i is an *interior* node (those within the convex hull).
    The order of the *interior* nodes is irrelevant to the routines.

Then, to generate all permitted tours to be examined by CheckThisPath, we make the single call:

    Visit(0,0,Gk)

In order to use these routines to find the TSP solution, CheckThisPath must compute the total distance using the particular order of node indices given in this path[], where those indices are as specified above. CheckThis-Path must retain the current shortest distance and its corresponding path at each call. Then the path with the "winning" minimum may be reported after the main call to Visit(0,0,Gk).

Note that the routines as given here have a *variable* complexity depending on the value of Gk and Gn. In fact, by our hypothesis it will run in $O(Gn^{Gk})$ time, hence caution is advised in using them on "difficult" cases, which depend on the size as discussed in the main text.

Additional pruning may be done by pre-computing a matrix indicating whether a given edge intersects any other edge, as exemplified in the second algorithm. The effectiveness of this method depends upon the growth of the number of tours-without-crossings with respect to the number of nodes in the graph, which is related to the number of edge intersections in the graph – but we will leave all this for our graduate students to explore.[54]

---

54 See Deut 25:4: "Thou shalt not muzzle the ox that treadeth out thy corn on the floor." Alas, I have no graduate students at present. Too few are willing to do the hard jobs, and even fewer are willing to endure the risks at my lab. Several Federal agencies (including the NSA, the NRC, and OSHA) keep threatening to close it due to the exceedingly high exposure to levity. Some Federal workers claim that they are unable to detect such radiation. This comes as no surprise; there are too many in academics and industry who suffer from the same defect. But then "...really one can do nothing beyond laughing, unless one dies of laughter." GKC *ILN* March 20 1909, CW28:293.

Routine Permute

The routine Permute generates the next permutation of the current partition of the interior nodes. It either calls itself if more interior nodes remain in this partition, or it calls Visit to resume the traversal of the convex hull, or if the tour is complete, it calls CheckThisPath.

```
routine Permute
;; NOTE this routine is only to be called by Visit
;; and is not for external use.

arguments
    in integer pathpos
    in integer nexthull
    in integer unfinishedinterior
    in integer thisgroup
    in integer doneinthisgroup
locals
    integer someinteriornode
begin
    ;; add the next permutation of a partition
    ;; containing (thisgroup) interior nodes

    for someinteriornode ← 1 to Gk
        if unused[someinteriornode] then
                unused[someinteriornode] ← FALSE
                ;; put this INTERIOR node into the path...
                thispath[pathpos] ← (Gh+(someinteriornode-1))
                if pathpos=Gn-1 then
                    ;; that was the last node to be put into the path
                    ;; so process it
                    CheckThisPath()
                else
                    if doneinthisgroup ≥ thisgroup  then
                            if nexthull < Gh-1 then
                                Visit(pathpos+1,
                                    nexthull+1,
                                    unfinishedinterior-thisgroup)
                            endif
                    else
                            Permute(pathpos+1,
                                    nexthull,
                                    unfinishedinterior,
                                    thisgroup,
                                    doneinthisgroup+1)
                    endif
                endif
                unused[someinteriornode] ← TRUE
        endif
    endfor
endproc
```

Routine Visit

The routine Visit adds the next node of the convex hull to the tour. It either calls Permute to generate a new partition of the interior nodes, or it calls itself to advance along the convex hull, or if the tour is complete, it calls CheckThisPath.

```
routine Visit
;; use: set Gn, Gh, Gk by the given graph
;; and initialize unused[] ← 1
;; then call Visit(0,0,Gk)

arguments
    in integer pathpos
    in integer nexthull
    in integer unfinishedinterior
locals
    integer thispartsize
begin

    ;; advance to the next node on the hull

    ;; put this HULL node into the path...
    thispath[pathpos] ← nexthull

    if pathpos = Gn-1 then
        ;; that was the last node to be put into the path
        ;; so process it
        CheckThisPath()
    else
        ;; add a permuted partition of the interior nodes...
        for thispartsize ← unfinishedinterior to 1 step -1
                Permute(pathpos+1,nexthull,unfinishedinterior,
                        thispartsize,1)
        endfor
        ;; then advance through the hull
        if nexthull < Gh-1 then
                Visit(pathpos+1,nexthull+1,unfinishedinterior)
        endif
    endif
endproc
```

APPENDIX 4: STIRLING NUMBERS OF THE FIRST KIND

In standard reference works the Stirling Numbers of the First Kind are defined in the following recurrence formula:

$$S1(p,q) = S1(p-1,q-1) - (p-1)S1(p-1,q)$$

with

$$S1(p,p) = 1$$
$$S1(p,0) = 0$$

Stirling Numbers of the First Kind: $S1(p,q)$

p\q	9	8	7	6	5	4	3	2	1
1	.	.	.	.	.	.	.	.	1
2	.	.	.	.	.	.	.	1	-1
3	.	.	.	.	.	.	1	-3	2
4	.	.	.	.	.	1	-6	11	-6
5	.	.	.	.	1	-10	35	-50	24
6	.	.	.	1	-15	85	-225	274	-120
7	.	.	1	-21	175	-735	1624	-1764	720
8	.	1	-28	322	-1960	6769	-13132	13068	-5040
9	1	-36	546	-4536	22449	-67284	118124	-109584	40320

These numbers provide the coefficients for the "semifactorial" (also called upper or descending factorial) polynomial of n terms:

$$x(x-1) \ldots (x-(n-1)) = \prod_{j=0}^{n-1}(x-j) = \sum_{j=1}^{n} S1(n,j)x^j = n!\binom{x}{n}$$

For example,

1: x $=$ $(1)x$

2: $x(x-1)$ $=$ $(1)x^2 + (-1)x$

3: $x(x-1)(x-2)$ $=$ $(1)x^3 + (-3)x^2 + (2)x$

4: $x(x-1)(x-2)(x-3)$ $=$ $(1)x^4 + (-6)x^3 + (11)x^2 + (-6)x$

5: $x(x-1)(x-2)(x-3)(x-4) = (1)x^5 + (-10)x^4 + (35)x^3 + (-50)x^2 + (24)x$

Rearranging this slightly to suit our reduced general form of the expected complexity of our method, we may write:

$$T(n,k) = (n-1)(n-2)\ldots(n-k) = \prod_{j=1}^{k}(n-j) = \sum_{j=0}^{k} S1(k+1, j+1)n^j$$

Thus we have

$T(n,1) = (n-1)$ $=$ $(1)n + (-1)$

$T(n,2) = (n-1)(n-2)$ $=$ $(1)n^2 + (-3)n + (2)$

$T(n,3) = (n-1)(n-2)(n-3)$ $=$ $(1)n^3 + (-6)n^2 + (11)n + (-6)$

$T(n,4) = (n-1)(n-2)(n-3)(n-4) =$ $(1)n^4 + (-10)n^3 + (35)n^2 + (-50)n + (24)$

BIBLIOGRAPHY

Note: All Bible quotes are from the Douay-Rheims version.

Abramowitz, Milton, and Stegun, Irene A. *Handbook of Mathematical Functions with Formulas, Graphs, and Mathematical Tables.* Applied Mathematics Series 55. (Washington, D.C.: National Bureau of Standards, 1972).

Aho, A. V., Hopcroft, J. E., and Ullman, J. D. *Data Structures and Algorithms.* (Reading, MA: Addison-Wesley, 1983).

Applegate, David L., Bixby, Robert E., Chvátal, Vašek, and Cook, William J. *The Travelling Salesman Problem.* (Princeton, NJ: Princeton University Press, 2006).

Chesterton, G. K. His collected works (CW) are published by Ignatius Press in San Francisco.

—. *The Everlasting Man.* (In CW2)

—. *Heretics.* (In CW1)

—. *Illustrated London News* essays (In CW27-36)

—. *Lunacy and Letters.*

—. *Orthodoxy.* (In CW1)

—. *The Poet and the Lunatics.* (In CW9)

—. *What's Wrong With the World.* (In CW4)

Čulik II, K. A Model for the Formal Definition of Programming Languages. *International Journal of Computer Mathematics*, Section A, Volume 3 (1973), 315-345.

Gellert, W., Küstner H., Hellwich, M., Kästner, H. (eds.) *The VNR Concise Encyclopedia of Mathematics.* (New York: Van Nostrand Reinhold Co. 1975).

Gerald, Curtis E. *Applied Numerical Analysis.* (Reading, MA: Addison-Wesley Publishing Co., 1978.).

Glasstone, Samuel. *Sourcebook on Atomic Energy.* (Princeton, NJ: D. Van Nostrand Company, 1950).

Gray, Henry. *Gray's Anatomy.* (New York: Bounty Books, 1877).

Gutin, Gregory and Punnen, Abraham P., eds. *The Traveling Salesman Problem and Its Variations.* (New York: Springer, 2007).

Hayes, William. *Project: Genius.* (New York: Atheneum, 1967).

Jaki, S. L. *The Absolute Beneath the Relative and Other Essays.* (Lanham, MD: University Press of America, 1988).

Knuth. Donald A. *The Art of Computer Programming.* 3 vols. (Reading, MA: Addison-Wesley, 1973).

Maycock, A. L. *The Man Who Was Orthodox.* A Selection from the Uncollected Writings of G. K. Chesterton. (London: Dennis Dobson, 1963).

McHugh, James A. *Algorithmic Graph Theory.* (Englewood Cliffs, NJ: Prentice-Hall, 1990).

Newman, James R., ed. *The World of Mathematics.* 4 vols. (New York: Simon and Schuster, 1956).

Preparata, Franco P., and Shamos, Michael Ian. *Computational Geometry: an Introduction* (New York: Springer-Verlag, 1985).

Roberts, Fred S. *Applied Combinatorics.* (Englewood Cliffs, NJ: Prentice-Hall, Inc., 1984).

Sedgewick, *Algorithms.* (Reading, MA: Addison-Wesley, 1983).

Stone, Harold S. *Discrete Mathematical Structures and Their Applications.* (Chicago: Science Research Associates, Inc., 1973).

Tarjan, Robert E. *Data Structures and Network Algorithms.* (Philadelphia: Society for Industrial and Applied Mathematics, 1983).

* * *

www.ingramcontent.com/pod-product-compliance
Lightning Source LLC
Chambersburg PA
CBHW071225050326
40689CB00011B/2467